RADICAL WAR

T0073930

MATTHEW FORD AND
ANDREW HOSKINS

Radical War

Data, Attention and Control in the Twenty-First Century

OXFORD
UNIVERSITY PRESS

Oxford University Press is a department of the
University of Oxford. It furthers the University's objective
of excellence in research, scholarship, and education
by publishing worldwide.

Oxford New York
Auckland Cape Town Dar es Salaam Hong Kong Karachi
Kuala Lumpur Madrid Melbourne Mexico City Nairobi
New Delhi Shanghai Taipei Toronto

With offices in
Argentina Austria Brazil Chile Czech Republic France Greece
Guatemala Hungary Italy Japan Poland Portugal Singapore
South Korea Switzerland Thailand Turkey Ukraine Vietnam

Oxford is a registered trade mark of Oxford University Press
in the UK and certain other countries.

Published in the United States of America by
Oxford University Press
198 Madison Avenue, New York, NY 10016

Copyright © Matthew Ford and Andrew Hoskins, 2022

All rights reserved. No part of this publication may be reproduced,
stored in a retrieval system, or transmitted, in any form or by any means,
without the prior permission in writing of Oxford University Press,
or as expressly permitted by law, by license, or under terms agreed with
the appropriate reproduction rights organization. Inquiries concerning
reproduction outside the scope of the above should be sent to the
Rights Department, Oxford University Press, at the address above.

You must not circulate this work in any other form
and you must impose this same condition on any acquirer.
Library of Congress Cataloging-in-Publication Data is available

ISBN: 9780197656549

Printed in Great Britain by Bell and Bain Ltd, Glasgow

CONTENTS

PART 3
CONTROL

LIST OF DIAGRAMS

ACKNOWLEDGEMENTS

Over the two years we took to write this book, a number of people very generously gave much of their time and knowledge to help us. In the first instance, we should thank Bob Evans at the UK Army Historical Branch. Bob introduced us to each other several years ago while we worked on different projects. Andrew was doing an Arts and Humanities Research Council project on military organisational memory and the digital archive at the Ministry of Defence and Matthew had been asked by the Army to find out why they had selected the SA80. Inevitably, Bob helped us see past the small beer and we got to thinking about the bigger picture.

In addition to Bob, many other people have been immensely helpful in the development of this book. Matthew was lucky to spend a month at the Naval Postgraduate School in Monterey courtesy of Dr Leo Blanken. Leo made it possible to attend a Department of Defense hackathon in San Diego. We've also been given a great deal of support from Professors Beatrice Heuser, Mike Rainsborough, Bill Allison, Neil Stott and Daniel Todman and Drs Kevin McSorley, Craig Whiteside, Jack McDonald, Jack Watling, Nisha Shah, Ian Garner and Jeff Michaels. Mike and Matthew disagree on pretty much every facet of contemporary politics, but we still found enough common academic ground to argue the points out in a way that we hope has made for a better book. Indeed, the plurality of academic views was important to us, precisely because of the contentious political and cultural moment.

We'd also like to thank former students Kevin McCann and Jan Tattenberg and are grateful for lively and important conversations with Guy Yeomans and Matthew Moore. We're also grateful to Merryn Walters for making sure we didn't obfuscate. Brigadier Dr Khashayar Sharifi, Lieutenant-Colonel Callum Lane, Dr Sandeep 'Frag' Mulgund and Lieutenant-Colonel Dave Lyle get mentions for their insightful thoughts and feedback.

We are also grateful to Dr William Merrin for sharing his always ahead of the field ideas and concepts on war and media.

A very big thank you must also go to Dr Philip W. Blood who gave much of his time and effort to help argue through some of the concepts we develop here.

Finally, a big thank you to Michael Dwyer for sticking with this project.

Distracted by three under six-year-olds, Andrew would like to thank Matthew F. for his patience during the months of lockdown. Andrew is also very grateful for the support of his wife Rachel, and for the mini-revolutions that are Jasper, Atticus and Loretta.

Matthew would like to thank his mum, Gillie, for helping out during holidays, and his daughter, Sarah, for being brilliant. Matthew must thank his wife, Sally, for putting up with him over the final months of writing.

Diagram 1: Mapping the New War Ecology

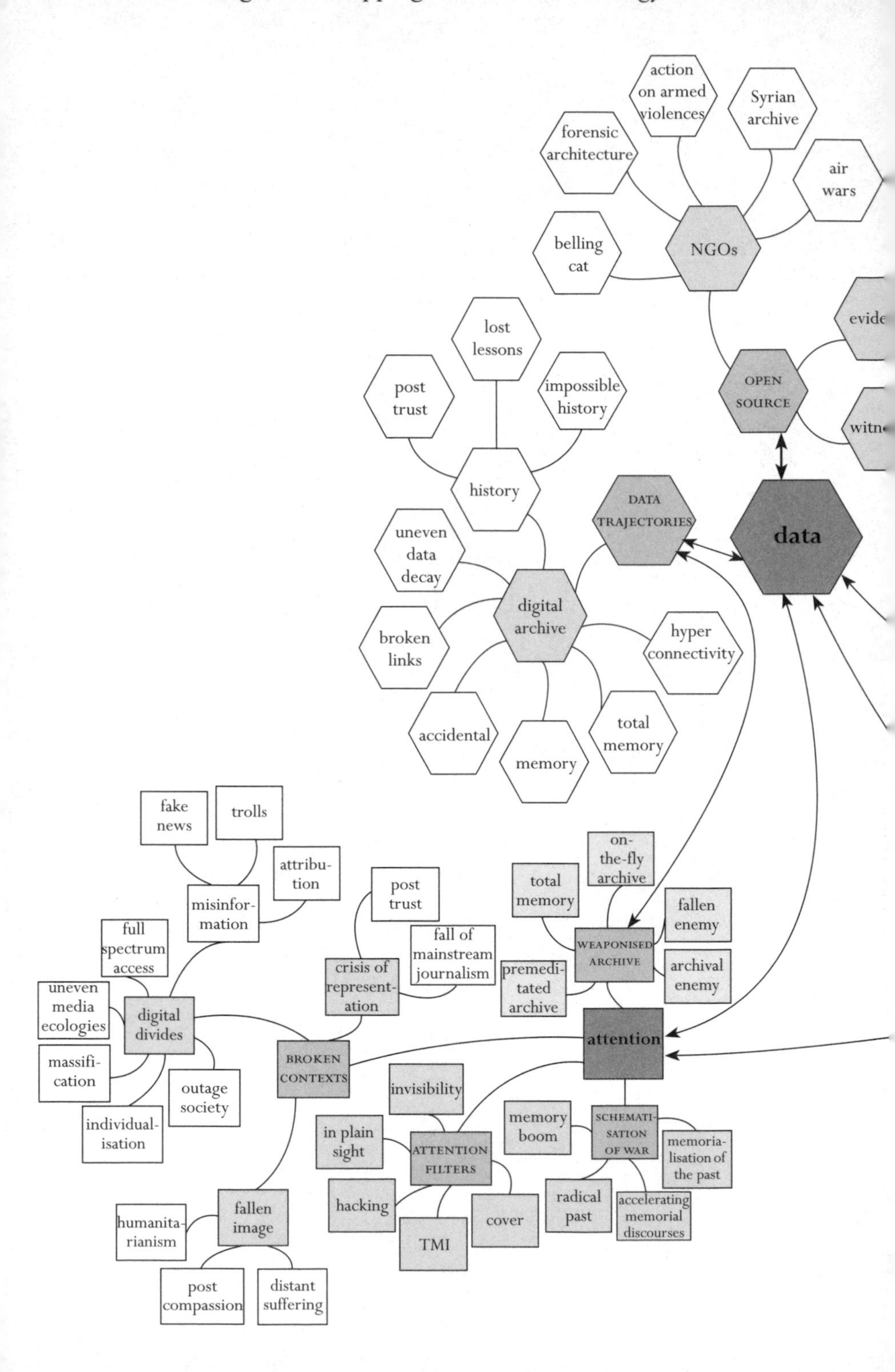

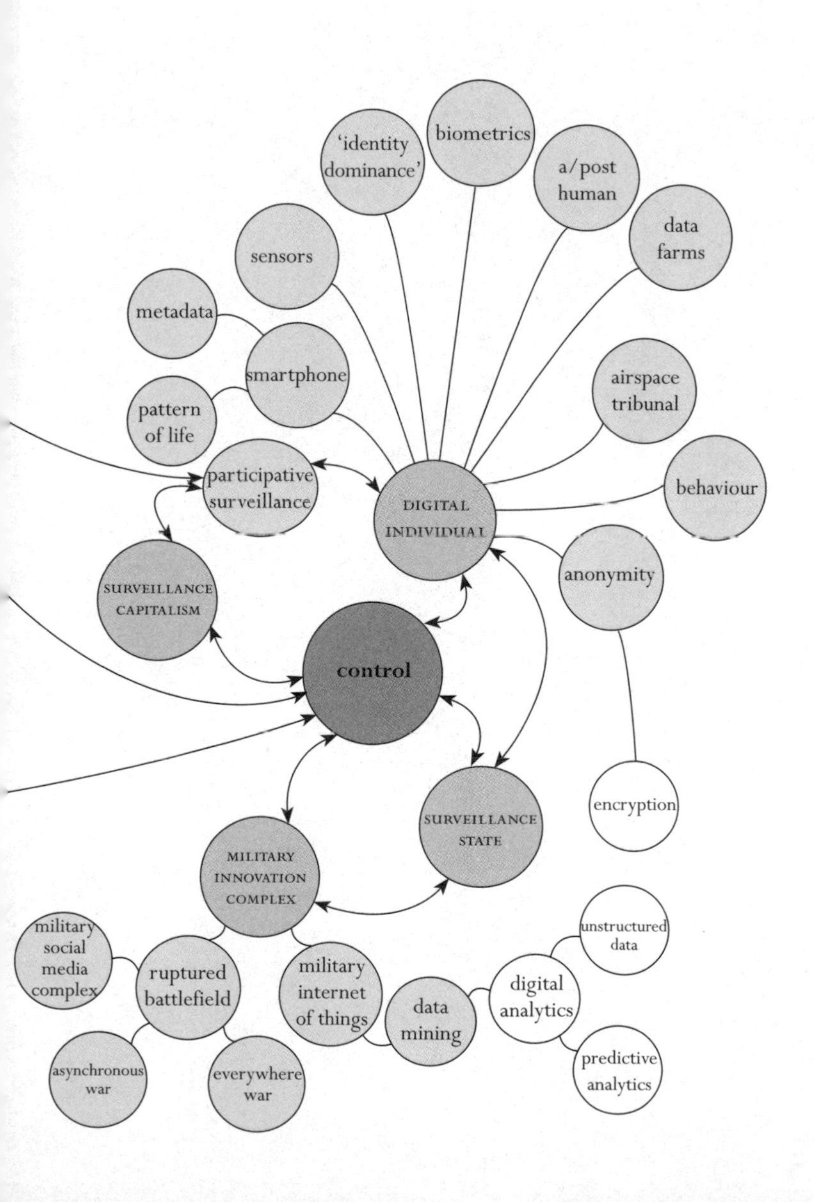

'identity dominance'
biometrics
a/post human
data farms
sensors
metadata
smartphone
pattern of life
airspace tribunal
behaviour
participative surveillance
DIGITAL INDIVIDUAL
anonymity
SURVEILLANCE CAPITALISM
control
encryption
SURVEILLANCE STATE
MILITARY INNOVATION COMPLEX
military social media complex
ruptured battlefield
military internet of things
data mining
digital analytics
unstructured data
predictive analytics
asynchronous war
everywhere war

PROLOGUE
WAR SINCE 9/11

A digital explosion is redefining the battlefield. It is reordering what we pay attention to and changing the way we understand war. In *Radical War*, we ask you to think again about war and media – particularly in the context of changes in how we consume media in the twenty-first century.

Defining what we mean by war takes us directly into an academic minefield. Since 2001, Wikipedia tells us that there have been 106 wars.[1] These include corporate wars, ethnic wars, counterinsurgency wars and drug wars. There have been wars of independence and identity wars; wars of genocide and wars for the purposes of the state. Some of the wars of the first two decades of the twenty-first century are household names. Others are lost in the peripheries of less common geography. British and American citizens are more aware of the Iraq War (2003–11), but how many of the leading Western news agencies have given prime time coverage to the Chadian Civil War (2005–10)?

While these wars have occurred all over the world, a substantial number have taken place in the Middle East, North, Central and West Africa, the Sahel, in Russia's near abroad, on the Asian subcontinent and around the Philippines.[2] Several of these wars have been reframed by the Global War on Terror (GWOT), and their protagonists became of interest to the United States. Others have an internal logic that has nothing to do with al-Qaeda – the adversary often cited as a key enemy.

All these wars constitute a subset of an expanding taxonomy in which the practice of war is no longer under the primary control of nations. As a globally accessible and editable resource that attracts over 21 billion page views per month,[3] Wikipedia has more traction in shaping debate than an academic might care to admit. Drawing from the 2014 Free Merriam-Webster dictionary, Wikipedia defines war as 'a state of usually open and declared armed hostile conflict between states or nations'. However, trying to make sense of the range of wars that have been fought over the past twenty years brings with it the need to expand the conceptual parameters with which we try to explain these events.

An example of this can be found in how Wikipedia applies its definition of war. The Kondovo Crisis (2004–5), for example, involving Macedonian police and Albanian rebels, is listed as a war. Yet, by Wikipedia's own Free Merriam-Webster definition this is not an 'open and declared armed hostile conflict between states or nations'.

Similarly, Wikipedia is at odds with itself regarding the ongoing Niger Delta (2004–) counter-independence 'war'. Here, the Nigerian government is supported by a number of countries, but nation-states are not the only protagonists. The delta's natural resources and the presence of Royal Dutch Shell have fuelled the conflict, with Shell being accused of financing militant groups as part of their efforts to retain control over the oil fields. In 2016, 15,000 militants surrendered, and their leaders were taken into custody to sign amnesty protocols. The death count in the Niger Delta remains vague.

If we are going to understand how these differing applications of the definition of war emerge, then we need to drill down into the factors that shape why these conflicts are defined as war. With regard to the conflict in the Niger Delta, for example, what we find is that there was barely any mainstream media (MSM) coverage in the West until 2006, when the BBC reported on it. Previously, the story had mainly been covered by African news agencies.

This poses some questions. What counts as war? Who was involved in writing up the Niger conflict in Wikipedia? How do they develop their understanding of what war is? And how does this reflect the perspectives of those who might be involved in and describe themselves as being at war? In *Radical War*, we view war as a form of political violence that involves 'the use of force to achieve a political end that is perpetrated to advance the position of a person or a group defined by their political position in society' (OECD 2016). Sometimes this form of violence is called war. Sometimes it is not. By treating war as a form of political violence, we direct attention to how we are made aware of conflict and how this has a bearing upon the way it is interpreted and internalised by societies.

Set in this context, it becomes easier to see why it is not possible to generate an accurate death count for the total number of people killed or injured in war in the twenty years since 2001. Far too often, conflicts rage where there is no infrastructure to record the casualties and consequences of contemporary wars. Death counts become contested by the belligerent and then politicised into dogma. In the end, no one speaks for the dead.

Instead, death counts and ratios of wounded to displaced persons are made sense of by a cohort of modern-day scholars, non-governmental organisations, international civil servants, foreign policy officials and members of the armed forces. Crucially, none of those protagonists access all the various forms of media in the same way. Their views of the facts have different perspectives, their insights are unevenly distributed.

Thus, for example, the Watson Institute at Brown University estimates that between 2001 and 2020 at least 800,000 civilians and military personnel have died as a direct result of military operations in Iraq, Afghanistan, Syria, Yemen and Pakistan.[4] The number of deaths from wounding or having fallen ill as a result of these conflicts is much higher still. Moreover, the Watson Institute calculates there are more than 21 million displaced persons from these wars, creating a refugee crisis that is greater than that experienced after the Second World War.

In *Radical War*, we map the informational, political and structural factors that explain why our knowledge and understanding of war is uneven. We look at the way war and its representation are constructed and disseminated, and how online and offline media interact to shape attention. Not everyone has access to the web, but the internet nonetheless influences our understanding of war in ways that are sometimes subtle or go unnoticed. We cannot quantitatively establish how media exaggerate or underplay the relationship between the fear of violence and its relationship to war and society. What we do know is that some wars are more openly reported on than others.

In this respect, the internet has become a vector for sharing knowledge about conflict, war and violence. According to the International Telecoms Union, an agency of the United Nations, 53.6 per cent of the world or 4 billion people were using the web in 2019.[5] According to Internet Live Stats, as of 26 November 2021 this has grown to over 5.1 billion internet users.[6] The ITU go on to observe that as of 2020, 96.7 per cent of the world's population can access a mobile-cellular network, 93.1 per cent can access a 3G network and 84.7 per cent can access a 4G network.[7] The ICT also establish that there is 99.3 per cent 3G network coverage in urban and 64.1 per cent in rural environments in Least Developed Countries (LDC) with 99.3 per cent of the population in LDCs having a mobile phone subscription (see Appendix 2).[8] Even more interestingly, Statista estimate that there are 6.3 billion smartphone users in the world in 2021, up from 3.6 billion in 2016.[9] Indeed, as of 25 November 2021, Internet Live Stats tells us that well over 4 million smartphones are sold per day.[10] Now connectivity is all about handheld devices.

This helps us understand how easy it is for people to access the web from different parts of the world. These figures also reveal that handheld technologies play a formative role in access to information. The world we live in is no longer conditioned by connections to fixed-line telecoms (only 11.9 per cent of the world's population connect to the internet this way).[11] Rather, our lives are framed and shaped by access to mobile devices and smartphones. Access to the internet is not evenly distributed. Digital divides remain around the world. These divides can be exploited. PR companies, governments, armed forces, intelligence agencies and a whole range of other actors can frame the ways in which we come to know about events taking place in different parts of the world. It is not possible to claim, however, that people's lives are not mediated by digital devices.

Thus, there are many reasons why some wars have registered in the public consciousness in some places but not in others. By 2021, however, the absence of MSM coverage cannot be considered a reason for ignorance. Wars are now being amplified by social media. Information – authoritative and unreliable; verified and unchecked – is compiled twenty-four hours a day, and then published and shared on platforms like YouTube, Facebook and Twitter. The public faces a constant media barrage of violence, coercion and killing. This moves seamlessly from the device in their hands, to their minds, to their memory, and is, for some at least, instantly forgotten with the simple swipe of a screen. For others, the very same media haunt and retraumatise victims and in the process prevent forgetting.

Radical War, then, is contemporary war that is legitimised, planned, fought, experienced, remembered and forgotten in a continuous and connected way, through digitally saturated fields of perception. Plotting the emerging relationship between data, attention and control, *Radical War* charts the complex digital and human interdependencies that sustain political violence today. Through a unique, interdisciplinary lens, this book maps our disjointed experiences of conflict and illuminates what we call the new ecology of war.

INTRODUCTION

WAR IN THE AGE OF THE SMARTPHONE

People's relationship to work and technology has changed enormously during the first two decades of the twenty-first century. The emblem of these changes is the iPhone and the generation of smartphones it spawned. Launched in 2007, these devices now make it possible to record events, find work, manage teams, locate ourselves on the planet, upload our experiences to social media, get a mortgage, read the newspaper, order a taxi, rent a holiday home, buy almost anything and get it delivered to your front door. The way that this device would revolutionise how people engaged with the internet, with each other and the world about them was not clear when it first came out. Slowly but surely, however, the smartphone and the platforms, services and applications that form part of the mobile, connected ecosystem have redefined the information infrastructures of the last century. These changes have generated huge amounts of data. This data can be mined for patterns and insights, revealing people's personal preferences and helping us to understand human behaviours. According to the World Economic Forum, these changes reflect the emergence of a fourth industrial revolution.[1]

The driving force behind this technology revolution are mainly, but not exclusively, high-tech entrepreneurs from Silicon Valley. The IT systems themselves had their origins in the US military's determination to secure communications in the face of a nuclear exchange. However, lacking the necessary finance and know-how to make more out of the systems that the military and US government had initially sponsored, venture capital firms backed a number of technology start-ups who would go on to help shape computing in the last few decades of the twentieth century. This made it possible to unlock the potential of the communication systems that the US government had initially financed.[2] The net effect of this was to establish a centre of technology innovation in the San Francisco Bay Area that has become the home of many entrepreneurs, financiers and technologists who now form a significant proportion of what we might call the virtual classes (O'Mara 2020).

These highly skilled technologists are not subject to production line labour practices. Instead, the virtual classes of the twenty-first century work over the web and form highly transient teams. Job insecurity is high, but the financial returns are worth it. Rewards are dependent on whether the virtual classes can disrupt existing business models and supplant people, material and machinery for web engagement, data analysis and prescriptive forecasting. This has been transformative for society and business. The convenience of using your computer to purchase something that had previously meant a trip to the mall revealed that consumers could be persuaded to use the internet to do more things online. As customers sought more opportunities to work while on the move, the smartphone ecosystem expanded. Now the smartphone powers the gig economy, managing workflow while monitoring the performance of a precariat workforce. The net effect has been to redraw how a whole range of businesses direct their logistics and automate their value chains, engage with their customers and manage their workers. Today, the smartphone has become 'the place where we live' (see Appendix 2 for connectivity data).[3]

These changes have established a services industry of lawyers, consultants and systems engineers who have driven and been the principal beneficiaries of globalisation. These people have implemented and managed the outsourcing and offshoring of business processes and manufacturing. The corresponding effect is that data has now hollowed out bricks and mortar, sometimes driving well-known businesses to failure even as others have released profits that are otherwise locked up in people, property and processes. In this manner, the professional services industry has helped sustain the competitiveness of established industries whose business models have been disrupted by the efforts of the virtual classes. When combined with low labour costs, industrial reform, admission to the World Trade Organization and careful use of free trade agreements, China has been particularly well placed to take advantage of changes in Western business practices. This has resulted in a process of economic rebalancing towards China, a process that is now reframing geopolitics in the early twenty-first century. These geopolitical changes may produce splinternets, or in the case of China, a decoupling from the rest of the internet (Inkster 2020). Irrespective of whether this happens, these changes have dramatically revealed the West's impotence in the face of Chinese security clampdowns in Hong Kong in 2019 and 2020.

The uneven effects of digitalisation and globalisation have affected various parts of society differently. This has been exacerbated by the failure of the banking system in 2008, which revealed and accentuated how some parts of society have been exploited more than others. In the decade after the financial crash, it also became more obvious how far supply chains have globalised. Local communities now recognise that they are at the beck and call of business decisions taken in distant parts of the world. At the same time, governments complicit in the liberalisation of the economic order have also struggled to demonstrate any responsiveness to the concerns of their citizens. In the United States and across several European countries, fiscal austerity following the 2008 financial crash pushed left-wing political parties further left. However, this was not enough to re-engage and mobilise an electorate who some believed had been betrayed by the political elite since at least the mid-1990s (Lasch 1996). In contrast to those on the left, populist politicians have been more successful in capturing the public imagination. Donald Trump, for example, has masterfully used Twitter to highlight the hypocrisy of the political elite, connecting directly with his audience and securing his electoral base by talking up social, cultural and economic divides. This new form of political discourse has de-emphasised policy and facts and instead reframed establishment politics in terms of self-interest and out of touch liberal elites.

Society's changed relationship to work has left people vulnerable to economic and political exploitation. The events of 9/11, the

GWOT, the 2008 financial crash, the Islamic State (IS), Russian and Chinese revanchism and now COVID-19 all represent a series of cascading geopolitical and geo-economic events that have been enabled and amplified by these new technologies. Government bureaucracies struggle to adapt to the changes heralded by digitalisation. The armed forces are structured for twentieth-century war even as they are being hollowed out by the same forces that are affecting businesses and the rest of society. At the same time, the smartphone ensures people can participate in work and conflict in ways that were previously inconceivable.

Our lack of a framework to make sense of these changes is evident when we look at social media. Designed to keep digital individuals coming back by creating a craving for hits and likes, social media flattens our appreciation for political violence, decontextualising it from its immediate social circumstances and reframing it through a global online community of like-minded followers. The June 2020 terror attack by a Libyan jihadist in Forbury Gardens in Reading, England, for example, was deeply traumatising for the families involved and those who were present. Three people were killed and a further three seriously injured. Though they might themselves have become victims of the attacker, people recorded the events on their smartphones and uploaded the films to YouTube.[4] Although the footage of the attacks was subsequently taken down from YouTube, this mass witnessing of violence had been amplified across a number of social media platforms. Politics has radicalised in ways that exploit anxieties and grievances as they are expressed on social media.

Similarly, when the Metropolitan Police shot dead a man who had strapped a fake bomb to his body on London Bridge in November 2019, a bystander recorded the incident and posted the actual shooting on Twitter.[5] Although the motivations, actors and the geographical locations were different, posting the material on social media reframed how these violent acts would be interpreted. Social media exposed the spectacle of the attacks and the police shooting to international audiences. The world could see what was happening in Britain. Immediately, these images were seen in a global and not just a local context. This decontextualised the images from the circumstances of the attacks. In the London Bridge attack, the attacker had purposefully strapped a false bomb to themselves in an attempt to signal that they were conducting a terror attack. However, the deliberate act of wearing a fake bomb also indicated that the attacker wanted to provoke the police into shooting him. This implied the bomber had a sense of police firearms protocols and could use this knowledge to lure firearms officers into martyring him through false signalling. As a result, the attacker was already appealing to their political base even as they understood they were alienating law-abiding British citizens. If you watched these events and were naïve about British politics since 9/11, then the most obvious feature of these videos was that people had been killed. The difference between the actions of the state and the terrorist had been levelled. New audiences had been given a ringside seat that enabled them to repeatedly watch, rewind, download, redistribute and discuss the meaning of these violent acts. All of this happened in ways that were well beyond the control of the state, opening up debate to audiences well beyond the UK.

The interaction between connected technologies and unusual acts of political violence has created new opportunities and challenges for the armed forces. From a military perspective, mobile connected devices are revolutionising how armed forces organise for warfighting. Cloud-enabled, networked and handheld digital systems are shaping the way the armed forces think through the necessary steps needed to go from collecting target information to taking a decision and dropping a bomb. This 'kill chain' (Brose 2020) has become open to automation, roboticisation and the application of Artificial Intelligence (AI). At the same time, some imagine a Military Internet of Things (MIOT) (see Appendix), where military sensors act as nodes for collecting data and thus the means by which targets can be rapidly identified. These systems of systems have the potential to remove people from the data collection and analysis process, and when combined with the delivery of ordnance by remote vehicle, to accelerate the way armed forces engage adversaries.

If all this could be made real, then armed forces might completely reimagine warfare and create the conditions in which they would no longer be dependent on maintaining large technological platforms. Organisations like the US Department of Defense (DoD) collect vast amounts of data. As Eric Schmidt observes, the AI systems that might help make sense of these data cannot be made to work without accessing commercial enterprise-wide cloud computing.[6] Inevitably, this means drawing on private sector expertise in the delivery of the sorts of data analytics the armed forces will need given their ambition to rethink the kill chain. According to this line of thinking, then, the MIOT represents an opportunity to reduce the overall footprint of the armed forces by stripping out unnecessary platforms. Just like commercial businesses, this suggests that the adoption of smart devices will lead large professional armies to redefine their roles. If this is right, then just as online shopping is helping to make the mall obsolete, the military must either face up to the prospect of significant structural change or the possibility of defeat in war (McFate 2019).

In conflict scenarios short of full-blown war, the logic of the MIOT extends out into civil society itself. In this respect, the civilian Internet of Things (IOT) (see Appendix) or mobile connected devices like the smartphone can also be weaponised. Like the MIOT, these devices are nodes in the kill chain, but they are also a way to spread disinformation. As a means for collecting intelligence, the smartphone makes it possible to track whom you talk to, where you go, whom you meet, how you travel and what you buy. What you witness and portray in the videos taken from your phone's camera helps to identify targets who can then be engaged kinetically or influenced politically. The smartphone thus becomes a vector for how the military conceptualise the future battlefield, shaping how they think about delivering ordnance to targets – like Amazon delivering parcels to customers.

The cumulative effect of these geopolitical, geo-economic and technological changes has altered the way that war is understood and conducted. War and its representation have collapsed into each other. Smart devices become both a way to represent war and a node in its practice. Media and weapon have enfolded each other. The information infrastructures that enable the production and consumption of data also enable the targeting of individuals in ways that have more in common with the manhunt than they do with conceptions of war as a duel. Smart devices offer up the possibility that defeat can be imposed without the enemy knowing they were being targeted. For the most part, the kill chain is hidden, only becoming observable at the point at which ordnance hits distant targets. This new war ecology ignores state boundaries and legal jurisdictions. Transgressing the binary distinctions between inside and outside the state has been a regular feature of political violence for centuries. The difference today is that no one can pretend that these transgressions are not happening.

As enabled by the smartphone, war is now participative in an evolving range of ways. People can now produce, publish and consume media on the same device. This fundamentally disrupts the control of the official narrative by the state's communications managers even as it replaces existing news platforms and channels. Now the state must co-opt the companies that manage the data flows that frame the experience of war or they must try to regulate them in ways that compel them to hand over data. In open societies, global technology companies find themselves in unique positions of power, capable of influencing national government decisions by threatening to remove services or investments. They can silence Donald Trump, who as outgoing president of the United States had to set up his own social media platform to regain his voice.[7] Similarly, when faced with regulation or taxation, Facebook have threatened to remove mainstream news content from their Australian platform and told the British government they would pull UK investment.[8] Set against this, in more authoritarian states the law may simply be rewritten in such a way as to force hi-tech businesses to choose where and how they operate. In both cases, however, the virtual classes have become the key to unlocking the state's capacity to regain control over the battlefields of the twenty-first century. At the same time, the state lacks the expertise or the knowledge to properly control the technology that it is now dependent on.

How war is experienced, recorded and understood has changed radically. Whereas once only soldiers and embedded journalists had privileged access to the battlefield, now war is everywhere, brought to us by the power of the smartphone and the information infrastructures it relies on. The consequence of this is that the companies that mediate our interaction with war and the world wield phenomenal power to shape the way in which war is communicated and experienced. This creates gaps between what governments and militaries say war is for, and what others perceive it to be. This has clear implications for the way that political violence is understood by those who are subject to it, suggesting that what the professionals of violence do and how that is experienced has ramifications in how force is justified.

In our efforts to scrutinise these phenomena, this book deconstructs the challenges presented by Radical War. In Chapter 1, we examine the confusing and opaque problem space that emerges out of the interaction between connected technologies, human participants and the politics of violence. Our aim is to examine war and its representation in the twenty-first century and to identify some of the central challenges that make up what we call a new ecology of war.

Chapter 2 sets out the analytical tools to help make sense of the changes that have been brought about through the emergence of the new war ecology. This chapter defines what we mean by hierarchies of violence and participation in war. This will help orientate readers to the implications Radical War poses across our three organising dimensions: data, attention and control.

Chapter 3 is focused on data trajectories and how data moves at different speeds depending on the information environments it occupies. We spell out the narrative implications of this along two main axes. The first is in relation to accelerating warfighting. The second relates to how military bureaucracies make sense of war. We argue that these two dynamics are incompatible with each other and are rupturing our understanding of war.

Chapter 4 shows how these data trajectories emerge into and shape popular discussions as they combine and fold with established narratives about war. This guarantees memory has more significance than history. In turn, we show how the schematisation of memory (refer to the Appendix) frames the way attention is being secured.

Chapter 5 considers the changing role of the digital archive (for our definition of an archive, see Appendix) as both a repository and the means for target identification through data mining. This sets out how attention is premediated through the technologies that are employed. It changes patterns of adversary identification from viewing enemies in terms of their iconic status to viewing them through the lens of the archive. The result is an infinite capacity to produce targets.

Finally, Chapter 6 is concerned with how control and influence work in the new war ecology. Here we consider the intersection of information infrastructures and the changing utility of military power. These infrastructures shape the capacity to influence decision-making both for the armed forces and for those individuals having to decide what networks to connect to even as they try to shape the online narratives.

We bring all our findings together in the conclusion and in the epilogue and encourage further engagement with the approach we set out here. In particular, our goal is to advance current understandings of war beyond narrow concerns with AI, machine learning and cyber-attacks. We do this by stimulating greater disciplinary engagement between those working in the fields of War Studies and those thinking about Media Studies. We hope all of this will help us to understand how we can come to know war under contemporary conditions.

Thus, *Radical War* encourages us to think again about the relationship between war and media in the context of twenty-first-century planetary-scale computational infrastructures (see Appendix). The relationship between war and society has been reconfigured by smart devices. These changes are now accelerating through an ongoing process of datafication (see Appendix). This plays out in how society and the state engage with each other, premediating our attention and challenging long-held assumptions about the relationship between the people and war, history and memory.

For the past two centuries, the Prussian philosopher of war, Carl von Clausewitz, has helped us understand how the French Revolution unlocked national passions and created a Revolution in Military Affairs (Paret 2004). This was not technology driven but emerged out of an interaction of social, economic and political factors (Howard 2001). Since 2001, however, the relationship between political violence, society and smart devices has changed how we come to understand

and make sense of war. In the twenty-first century, the smartphone makes it possible to mobilise populations, replacing the rifle as the weapon of choice for those engaged in mass participation in war. Smart devices, apps, archives and algorithms remove the bystander from war. This collapses hierarchies to a point where we are either victims or perpetrators. The distinctions between audience and actor, soldier and civilian, media and weapon become meaningless. In this context, Clausewitz's trinity of state, people and armed forces becomes irrelevant. Clausewitz cannot help us explore how this has happened. Today, we live in the fallout of this post-Clausewitzian age of war, an age we have identified as Radical War.

Of course, there are plenty of historical antecedents that predate the arguments cited in this book. However, our findings lead us to conclude that we need to rewrite how we come to know and understand war. The theories of war that we have grown familiar with can only take us so far. We do not argue that the state's capacity to generate violence is unimportant. Rather we want to move beyond state-centric interpretations, towards a framework that takes into account the changed relationship between technology, participation and war.

Contemporary war is a phenomenon that is being made sense of through digital, connected devices. This does not constitute a separate domain but is integral to how we understand war itself. In *Radical War*, we do not present a complete theory of war in the twenty-first century. We use the language of now because we cannot chart the future. The model presented here helps to map the contours of the disjointed experience of war in the expectation that our framework will contribute to an ongoing discussion and debate. We would like to continue and be a part of that discussion online at www.radicalwar.com.

RADICAL WAR
A DEFINITION

Radical War is the immediate and ongoing interaction between connected technologies, human participants and the politics of violence.

Political violence involves 'the use of force to achieve a political end that is perpetrated to advance the position of a person or a group defined by their political position in society' (OECD 2016).

This form of war is no longer primarily at the behest and control of nation-states. The principal actors include technologists, the professionals of violence, non-state actors, corporate organisations and the connected mass of users of digital and social media. War is no longer about compelling enemies to do the will of the state. Now war is principally about managing the attention of populations and different audiences where the will of the public is a constantly churning spectacle of opinions and perceptions that spill out and feedback into each other, irrespective of whether they are expressed online or not.

Radical War is 'radical' in that it distorts perceptions of the relationship between media, the military and war's political effects. Radical War is always seen through a prism of information infrastructures that are themselves opaque and poorly understood, even by those who have built them. Radical War is 'war' in that it is concerned with political violence; often with unforeseen outcomes. Identifying how or why violence occurs is masked by the systems that make it possible.

Radical War is fought in a battle and information space that we call the new war ecology (see Appendix). The new war ecology is an environment in which mobile, connected devices like the smartphone enable digital individuals (see Appendix) to share and create content that can influence politics and produce lethal effects. These digital individuals may be representatives of the state but can also act independently of governments, the military and MSM.

Fixed broadband connectivity exists in some countries and not in others. Some media ecosystems rely on social media while others do not. This produces multiple new ecologies of war, all with their own constraints and enablers, user experiences and political, military and societal dynamics. Even though some people do not have access to the internet, these new war ecologies still have effects on our perception and understanding of war.

Digital individuals may willingly participate in war or they may participate by dint of being connected to the grid. Radical War is participative in that everyone has the potential to be involved through the data they create. This removes the bystander from war and collapses the relationship between audience and actor, soldier and civilian, media and weapon.

Radical War compresses the experience of war into an intense spectacle in which local, national and transnational narratives and identities are brought into new conflict. This is intensified and enabled through an astonishing agitation of the history and memory of past wars, many of which are now accessible via social media archives and search engines. In this so-called 'post-trust' environment, shared realities are fragmented and counter-narratives manufacture doubt, uncertainty and conspiracy theory.

Radical War removes and reduces the boundaries of conflict, flattening our experiences and saturating our senses. By collapsing actors and audiences to the same level of participant, Radical War establishes a hierarchy of violence (see Appendix) between those who actively and those who unwittingly participate. These hierarchies ensure that people facilitate violence irrespective of their political preferences or personal choices. This contributes to the creation of Radical War, where mass participation bleeds into a war of all against all.

1

WAR AND THE DEMOCRATISATION OF PERCEPTION

Gamers were all fingers and thumbs on their gamepads. They were watching livestreamed games on Twitch, a video-sharing site owned by Amazon and designed for e-gamers. Social media was buzzing. Facebook had the scoop. What was Twitch livestreaming? A Neo-Nazi terrorist attack against two New Zealand mosques. The gunman self-streamed the deed. They hoped to kill as many as they could.[1] In seventeen minutes, there were fifty-one dead and fifty injured. This was Christchurch, New Zealand, on 15 March 2019, and the world saw everything. The *New York Times* tried to scoop the headline – 'A Mass Murder of, and for, the Internet' – but the gunman had out-scooped a global news brand.[2] The incident was posted, reposted and posted again in successive waves of sharing across YouTube, Twitter and Reddit. Then a sense of morality kicked in, with platform moderators struggling to remove horrific clips as a race ensued against those masses posting and reposting. Details emerged from the attack: the killer shouted a meme, 'subscribe to PewDiePie' – a reference to Felix Kjellberg, a Swedish digital games player. Kjellberg owned one of the most subscribed YouTube channels and had been accused of having links to anti-Semitism and alt-right neo-Nazi movements.[3] The social media multiplier was spinning beyond control. Facebook's Newsroom tweeted: 'In the first 24 hours we removed 1.5 million videos of the attack globally, of which over 1.2 million were blocked at upload.'[4] Reddit acted to ban any forums named 'gore' and 'watchpeopledie', forums that had acquired 300,000 subscribers within hours.[5] Forum moderators argued the video should be kept open as it offered 'unfiltered reality', but Reddit's content controllers faced a tidal wave of information to try and manage, and in the end the pages came down.[6] By that point, it was impossible to gauge the scale or intensity of shared viewing on WhatsApp, Telegram and other encrypted messaging services.

Livestreaming and reproduction of violent political imagery is an everyday event. This radicalising content is the real-time battlefield, reshaping discussion irrespective of whether someone owns a smartphone or not. It is also a global game between those who post violent images and the guardians who try to take them down – it is live, it is online, and it is a race. The complexity of digital infrastructures that underpin this arena remove any certainty that all the images of the Christchurch attack were removed. The scale of uncertainty is amplified by the opacity of the information infrastructures themselves. Consequently, we cannot know how many copies of the video there were or how many viewed the attack live. The number of links sharing the content and the quantity of narrated commentary remain unknown.

In response, Facebook and other social media platforms established the Global Internet Forum to Counter Terrorism (GIFCT) to try to help prevent this sort of livestreaming in the future. As the copycat attack by Stephen Balliet against the Jewish community in Halle, Germany, in October 2019 demonstrated, GIFCT proved inadequate to the task almost from the get-go.[7] Balliet, as the actor-killer, issued an online manifesto for distribution via an alt-right extremist message board 8chan before livestreaming the attack on Facebook and Twitch. Real life now mirrored a first-person shooter game. Around 2,200 people viewed Twitch before the stream was pulled down. The normal content would otherwise be friends participating in video games. This new dystopian show was all about two people killed and one injured.[8]

* * *

Violence is ubiquitous, some might even say mundane. People are subject to it in all sorts of contexts, and across all parts of society. This complicates understanding and underlines how violence is both contentious and subject to multiple interpretations (Miller 2020, pp. 5–8). Violence is represented in all sorts of media. The actuality of it even gets streamed live over gamer channels like Twitch. It can be domestic, hidden, religious, criminal or political. Violence orders social groups in different ways. In the military, it is about the controlled application of violence. In gangs, it might be designed to conjure fear to shape relations within and between rival groups. At a football match, violence might frame how fans identify themselves or provide opportunities to let off steam. For the media, violence creates clicks and drives attention. Within the state, the government claims a monopoly on the legitimate use of violence. Between states, political violence is war.

Today, our understanding of war and violence is mediated by connected technologies like the smartphone. These devices now saturate our experience of the world. Violence as it appears in these contexts produces intense spectacle in which local, national and transnational narratives and identities are brought into new conflict. Thus, the attacks in Christchurch inspired copycat attacks in Halle. The attackers targeted different communities, but their anti-Semitic and racist intent and their livestreaming through Twitch had both national and transnational political intent. How local and global audiences made sense of the violence and how the attacks produced political effects is nevertheless always contextual. However, if we accept that there is no going back to a pre-digital age, that no-one is going to turn off the internet, then there can be no political violence without its digital representation. In these circumstances, connected technologies like the smartphone help to create asynchronous experiences of war and violence. This produces a collapse of context and leaves audiences stuck in the moment, free to define the meaning of these experiences in whatever way they choose (Brandtzaeg and Lüders 2018).

In the twentieth century, states prosecuted wars and the media reported on them. However, in this century information infrastructures have created the conditions in which media is now made for the prosecution of war. This is a direct consequence of a convergence between the weapons of war and the media of war, between the means through which wars are fought and the means through which war is experienced. People participate in their own surveillance – using smartphone apps, uploading geolocated video – and in the process, we contend, they have become part of the machinery of war, enabling the delivery of ordnance to targets. This, above all else, demands a re-examination of how war in a deeply mediated world now works. For the explosion of data and smart devices has weaponised attention, de-territorialised war and made the seizure of minds both easier and cheap.

For example, WhatsApp is an end-to-end encrypted messenger service owned by Facebook. Free to download to your smartphone, WhatsApp can be used to prosecute war overseas and organise political violence at home. Overseas, WhatsApp was in use among armed forces coordinating Reaper drone attacks in Mosul.[9] American forces have been advised to download the app for operational use on their phones,[10] and it has been hacked by Israeli 'cyber-arms dealer', NSO Group. While the 2019 vulnerability was patched by WhatsApp, NSO Group has gone on to extend its Pegasus software to collect data from phones without even having to work through an app.[11] The result of all this is that the smartphone becomes a sensor. It can 'collect intimate data from a target device, including capturing data through the microphone and camera, and gathering location data'.[12] This is not just about the United States or the NSO Group either. In Ukraine, the Russian government uses Telegram, the cross-platform, instant messaging system, to keep track of people and promote division.[13]

Compare this with the way that WhatsApp, Instagram and social media sites like Parler and Gab were used by supporters of President Trump to organise an insurrection and storm the Capitol Building on 6 January 2021. Including veterans from the wars in Iraq and Afghanistan – one of whom, Ashli Babbitt, was shot dead by Capitol Police[14] – the insurrection's goal was to stop Congress from certifying President Biden's election victory. Recording and broadcasting events from their smartphones, the protagonists produced data that made it easy for the FBI to identify and then subsequently arrest them. At the same time, the events on the Capitol have created a digital archive for Trump supporters to look back on and invoke in their ongoing efforts to re-elect the forty-fifth president. The smartphone and the digital ecosystem it fostered have created all number of entirely new media for war and violence to occupy. In the process, we now experience a constantly churning spectacle of opinions and perceptions that spill out and feed back into each other, irrespective of whether they are expressed overseas or at home.

For the evangelists of the information age, the internet would lead to the spread of democracy and the onward march of progress. In 2011, for example, networked technologies made it possible to amass large numbers of protestors in Tahir Square in Cairo; those protestors then called for the downfall of the Egyptian President Hosni Mubarak (Tufecki 2017). But for authoritarians, insurgents and terrorists, the twenty-first-century information ecology also lends itself to fostering division, fear and uncertainty. For instance, Afghans learnt to post whatever they wanted to Facebook, to store whatever information they liked on their personal devices and to leave images or data in the cloud without having to fear for the consequences. Since the Taliban retook Kabul in August 2021, however, Afghans have had to weigh up what parts of their digital selves they must delete for fear of retribution. In this case, '[t]he challenge is how do you balance getting information – like what's going on at the airport, and people trying to reach you – with eliminating evidence that a group would use to implicate you in something and take you round back to make an example of you'.[15] Put this way, the digital environment is not just another channel for distributing distrust among adversaries and cementing support among the converted but is the warzone itself. Now, online and offline experiences integrate, shape perception and drive people to action.

Information infrastructures exacerbate the distribution of distrust in ways that are uneven and with second- and third-order effects that are difficult to control. Regional wars have always offered an opportunity for larger powers to test new equipment and the associated concept of military operations. The Armenia–Azerbaijan war in September 2020, for example, made it possible for Israel to establish the effectiveness of their Harpy and Harop loitering munitions.[16] Employed by Azerbaijan's armed forces, the drones provoked much discussion on social media as to whether they had effectively brought the era of the tank to an end.[17] Much of this debate was nonetheless shaped by the availability of online media about these wars. Before the outbreak of hostilities, the Azerbaijani Army released a rock track called 'Fire' by Narmin Kerimbeyova, Ceyhun Zeynalov and the Nur group that heavily featured the Harpy drone in the music video.[18] As the war unfolded, this video went viral, shaping perception and framing the propaganda and social media commentary that followed.

The effectiveness of the drone aside, this conflict also offered an opportunity to test how state-directed information operations might be used on the battlefield. In particular, how might the different information infrastructures in Armenia and Azerbaijan shape the way that the conflict was portrayed? According to the non-governmental organisation Freedom House, Armenia has a more open approach to accessing the internet when compared to Azerbaijan.[19] This openness, however, could be used against Armenia. Indeed, Azerbaijan's capacity to dominate social media almost certainly had more to do with Baku's decision to employ cyber-attacks on Armenia while cracking down on its critics at home and directly taking control of its online propaganda efforts.[20] This swamped Armenia's more open approach to net freedom, leaving the country vulnerable to Distributed Denial of Service attacks that left many Turkish and Azerbaijani websites inaccessible and creating problems accessing TikTok.[21]

Considering the chaos these sorts of attacks might create, intelligence agencies have inevitably spent considerable time trying to understand how to adjust to the way civilian information infrastructures have radically refashioned social engagement.[22] The fear is that these new technologies will foster 'an embedded fifth column, where everyone, unbeknownst to him or her, is behaving according to the plans of one of our competitors'.[23] This has a military dimension, one in which confusion at home can be used to gain strategic or tactical advantage in war (Chotikul 1986; Thomas 2004).[24] One way that the armed forces have sought to get inside the challenges posed by subversion has been to try to accelerate the speed at which they make decisions. This would allow them to operate more quickly than those trying to undermine cohesion at home. For Western armed forces, this has meant integrating existing government structures, doctrines and digital platforms with those units responsible for carrying out air, land, sea, cyber and information operations.[25] In the UK, the hope is that these discrete spheres of military and government activity can then be fused to deliver coherent political effects at speed to gain 'information advantage'.[26] Nevertheless, it is not clear how we might realistically imagine 'information advantage' when hyperconnectivity, information overload, ubiquitous surveillance and rolling social media are contributing to a post-trust environment.

Seen through the prism of these bureaucratic structures, the experience of war remains framed by how the armed forces organise themselves. Looked at from outside the prism of the military, however, a useful way to think about experience under digital conditions is to talk about a 'dimension'. For example, Laurence Scott argues that digital technologies are reshaping what it is to be human to the extent that we are now 'four dimensional'. He argues that

> the fourth dimension doesn't sit neatly above or on the other side of things. It isn't an attic extension. Rather, it contorts the old dimensions. And so it is with digitization, which is no longer a space in and out of which we clamber, via the phone lines. The old world itself has taken on, in its essence, a four-dimensionality. (Scott 2015, p. xv)

Even if you don't have a Facebook account, the company boasts 2.91 billion active users who do.[27] In these circumstances, and given the levels of global connectivity, our subliminal day-to-day engagement with the world is entirely mediated by digital experiences. This is true whether users are recording a moment and uploading it to Facebook or organising a social gathering – even with people who do not use social media – via their smartphone. In our terms, this is a new and fertile space in which war is flourishing. And yet the military continue to try to make sense of these experiences through bureaucratic prisms as expressed in their organisation structures and doctrines. While this makes it possible to sustain a way of seeing that privileges the military's version of reality, sustains outmoded martial cultures and maintains doctrinal purity, it does not help to make sense of war as it might be understood by society more broadly.

Thus, in an effort to escape these military constructs, we explicitly reject Clausewitzian definitions of war. The use of violence is not exclusively under the control of the state or the military. Strategists and the military may prefer to define war as a continuation of politics by other means. By contrast, in Radical War, we seek to understand how political violence gains meaning in a 24/7 always online environment. By taking this broader approach, we can investigate the way knowledge about war has become a battle for control over the relationship between data and attention. Our analysis consequently decentres the battlefield, directs our principal concerns away from military strategy, and instead moves us towards a theory of knowledge about how we can know war in a networked, highly mediated world.

To replace the rigidity of the Clausewitzian model of war, we propose to map political violence across three dimensions that we label, data, attention and control:

1. Data: involves the intense connectivity and the datafication (see Appendix) of battle. This produces immense data volumes that enable multiple, simultaneous, messy and weaponised data trajectories, creating accidental archives and new human–machine configurations of perception. In these contexts, data can be any kind of information, so long as it's expressed in digital form. These shape how wars are used as 'lessons' and go on to (de)legitimise current/future strategy. This creates an:
2. Attention disorder that clutters and confuses what Paul Virilio (2009) describes as the grey ecology as it reshapes the relationship between knowledge, understanding and the battlefield. This creates a crisis of representation that fragments interpretations, further undermining the veracity and effect of narratives in a post-trust environment. This in turn frames the archival fight (mining and ownership) over memory and history, opening up space for artistic and citizen interventions that

challenge and reframe how we think of war and its relationship to human/legal rights. This both enables and requires new forms of:

3. Control that demand new information infrastructures (see Appendix) and techniques of surveillance. These influence investments and create new modes of situational awareness that shape and manage the participant's experience of battle, exploiting smart devices over new battlefields. Stresses in relation to combat and how participants engage online challenge the way we frame expertise and the production of knowledge about war in twenty-first-century societies.

In the rest of this chapter, we examine the confusing and opaque problem space that emerges out of the interaction between connected technologies, human participants and the politics of violence. Our aim is to examine war and its representation in the twenty-first century and to identify some of the central challenges that make up what we call a new war ecology. We do this in relation to data, attention and control, Radical War's central organising dimensions.

Data

Time is the multiplier of data. At any one minute in 2020, Zoom hosted 208,333 people, YouTube added 500 hours of video and Instagram posted 347,222 stories.[28] Radical War is only made possible by the information infrastructures upon which it depends. As the example of the Christchurch mosque attacks at the beginning of this chapter demonstrates, these infrastructures are ubiquitous and work at a scale and speed that had previously not been possible. As a compound of people, knowledge, processes, organisations and technical systems, information infrastructures are built out of 'facilities and services usually associated with the internet' and include 'systems that process and transport data inside and outside national boundaries' (Bowker et al. 2010). These information infrastructures now extend from our places of work to the IOT. This has led to a process of 'datafication' in which all aspects of life have been turned into online quantifiable datapoints. These are subject to new forms of power, exercisable by 'those with access to databases, processing power, and data-mining expertise' (Andrejevic 2014, p. 1676; Livingstone 2019, p. 171).

Datafication emerges out of processes of participatory surveillance (see Appendix). If someone owns a smartphone or a connected device, they are asked to register it with their service provider. This instantly creates a profile that constitutes the first step in becoming a digital individual. People can create multiple accounts. Consequently, it is perfectly possible for there to be more digital individuals than actual people in the world. People willingly become digital individuals because they gain access to goods and services that otherwise would not be available to them. The trade-off comes in the exchange of information people provide about themselves in terms of their location, search history, identity, sexuality, contacts, personal relationships and so on. This facilitates new informational trajectories that have reshaped contemporary power relations between people, technology and their experience of work.

Given that these informational trajectories are largely invisible, it is difficult as well as time-consuming to fully grasp the kinds and quantities of personal information that are collected, stored and shared about our daily lives. Sharing data brings many benefits. Recovering control over that data and reasserting a digital individual's personal privacy risks losing access to forms of digital communication and everyday services that are now taken for granted. This is because those who collect, manage and analyse data are unlikely to give up their place in the market and, if compelled to do so in legislation, may be more inclined to move their data centres to locations beyond regulatory control. There has thus developed a widespread state of resignation to the wholesale datafication of the self, to push privacy concerns to one side for the sake of convenience.

In parallel, technologists are designing devices and web platforms specifically to capture attention and keep people engaged in online environments. Social media, for example, is designed in such a way as to create opportunities for exchanges to become heated and emotional. This helps to polarise political discourse and can harm people's mental health. Indeed, as the Facebook whistle-blower Frances Haugen observes, 'Facebook is making hate worse.'[29] Engaging the senses is the explicit ambition of the software engineer.

Social media platforms have to be sticky. They have to feed our senses and encourage us to commit more of ourselves to the digital world than we might otherwise with our friends or families. Thus, social media offers both a vector for communication and the means by which participants indulge in their senses even as they provide key tracking information that can be exploited for business usage and military effect.

In this respect, war and media have meshed with everyday experiences. As Andrew Hoskins and Ben O'Loughlin wrote more than ten years ago, 'the planning, waging and consequences of warfare do not reside outside of the media' (Hoskins and O'Loughlin 2010, p. 5). And yet the intervening years of transformations in media and communication content, infrastructures and practices are where we have seen the emergence of Radical War. Paradoxically, the net result of this decade of change in media and communication is both a clarification and a distortion of our appreciation of war. On the one hand, the internet makes available more information than we ever knew we needed about war. On the other, even as new technologies enable connectivity, contemporary experiences are now so saturated with data that is it nigh on impossible to make sense of it all (Weinberger 2011).

Too much information is not a new phenomenon. There have always been a range of social filters for dealing with the challenge. These come in the form of personal networks that help frame what becomes known and what is ignored. In online environments, however, the filters that sift between wanted and unwanted information operate through menu choices and tick box selections. These trigger algorithms that filter forwards, bringing wanted results to the front of an online search while leaving other datapoints available but left in the background (Weinberger 2011).

It is not just that connected devices like the smartphone act as hyperconnected digital trackers. Rather it is also that they create an archive of past activity that is a constantly available resource for states and the companies that own and operate today's digital infrastructures and networks. That these are mobilisable in a moment was spectacularly demonstrated during the COVID-19 global pandemic. Smartphone data was used to build and mine databases, often in the name of 'self-diagnosis' or monitoring the movements of known patients to help control the spread of the virus. For example, as cyber-security specialist Zak Doffman explains:

> Our phones are the front-end to an absolute wealth of data that can all be collated and mined. Your phone knows where you go, yes, but it also knows who you know, who you speak to, how often. If I add the phone's location to details of who you know and look at where those people might also be, I create a dataset that provides a map of likely contact. And likely contact means possible infections. And I can go much further, building and manipulating that dataset using just metadata.[30]

In China, for instance, mass surveillance tools were used both to suppress anger over the government's handling of the COVID-19 crisis and to enforce quarantine measures.[31] But it is the latter, the containing of the outbreak, that provides an excuse for accelerating the mass collection of personal data, as part of a data grab to track individuals that could easily become the norm once the crisis is over. In Israel, technologies usually deployed for counter-terrorism, tracking citizens by geolocating their mobile phones, are being used for quarantine enforcement, to check the movements of those who are infected and those in the vicinity of the infected.[32] Meanwhile, the US Centers for Disease Control and Prevention (CDC) have turned to data provided by the mobile advertising industry, through apps such as maps and weather services, to analyse population movements in the midst of the pandemic. Whereas in the UK, location trend analysis to help combat the crisis employs anonymised and aggregated data, in the US, the CDC has acquired data that is pseudonymised, but not aggregated.[33] As data scientist Dr Yves-Alexandre de Montjoye explains:

> The original data is pseudonymised, yet it is quite easy to reidentify someone. Knowing where someone was is enough to reidentify them 95% of the time, using mobile phone data. So there's the privacy concern: you need to process the pseudonymised data, but the pseudonymised data can be reidentified.[34]

Thus, the digital archive has become the epicentre of Radical War, the place where individuals are folded into a myriad of potentially unlimited data manipulations. The key dimension of this 'unlimited' vulnerability is time itself, as it is the potential for meaning to be made about data that gives us war's radical uncertainty. The ease with which millions of people from countries across the world have been placed under smartphone-enabled surveillance in the name of enforcing quarantine rules or understanding the spread of COVID-19 is striking. This shows how participatory surveillance is anchored in an encroaching datafication of the mundane: the more habituated smartphone use becomes, the less averse people are to the threats posed by digital living. However, the data gathered and gatherable from participatory surveillance has potentially unlimited future uses in identifying and targeting individuals. This contributes to the accumulating risks from and anxieties that are engendered by Radical War.

War is no exception. The MIOT connects ordinary battlefield devices to a network that transmits data about and across the battlespace to help warfighters gain situational awareness. These battlefield devices reconfigure the nature and scope of the battlefield, meshing with civilian smart devices to reframe how we come to see, know and understand war. This creates more attack surfaces for the armed forces to act upon. At the same time, the process of datafication refashions the place and power of elite actors (militaries, governments, news media) and their relationship to participants, audiences, victims and perpetrators. Datafication has already had an impact on how wars are fought and experienced. As our analysis takes us further away from the uncertainty of the kinetic battlefield, we will show how these fast-changing and everyday digital cultures are reproducing uncertainties in how we think about archives, history, memory, civil–military relations and the conduct of war.

Attention

Overwhelmed by the data explosion, people's knowledge of the contemporary battlefield has fragmented in ways that cannot be filtered by the armed forces alone. Soldiers use smart devices connected to civilian cell phone networks to send messages home via social media and in the process reveal where they are located. Civilians stranded between battlefronts unintentionally connect to wi-fi networks owned by oppressive regimes and provide data and insights that can be used for targeting. This has implications for understanding the production, veracity and ownership of data, and yet these issues rarely form part of an analysis of how to make sense of contemporary war. In this regard, war is now radical. It is beyond the grasp of a collective consciousness. The churn of information that is now available online calls into question whether any kind of consensus on the waging, consequences and possible modes of preventing war is even possible. Thus, our argument runs counter to the idea that more information means better democracy and decision-making.

What makes it so hard to make sense of the new ordering of social relations brought about by participatory surveillance is that images of war and violence have no immediately discernible or linear pattern. Stories emerge asynchronously into our online worlds and cross fertilise; new meanings can then be constructed out of this new context. In this respect, the saturation and connectivity of data and information break and reframe experiences in ways that only serve to distort the what, when, why and who of war. The effects of this on how war is justified, explained and military power is used have hardly begun to be seen. Of course, this is not a particularly new phenomenon. People's views on war had already fragmented into a number of realities in the first six months of 2003, way before social media had made any sort of impact on twenty-first-century living.

On 15 February 2003, for example, a day of anti-Iraq War protests occurred around the world. Across more than 600 cities, between 10 and 15 million people marched against the US-led invasion of Iraq.[35] The French political scientist Professor Dominique Reynié estimated that 36 million people took part in nearly 3,000 protests between January and April 2003.[36] More people protested than there were inhabitants of Iraq.[37] This was the largest worldwide protest in history, and yet the United States and its allies ignored the protestors and invaded Iraq on 20 March 2003. Around six weeks later, President George W. Bush announced the end of the successful invasion in his

'Mission Accomplished' speech aboard the USS *Abraham Lincoln*.[38] As chaos broke out on the streets of Baghdad, Mosul and Fallujah, the idea that the war was over was an utter surprise to liberated Iraqis.[39]

The Iraq War was contentious and produced several different perspectives on its legitimacy. In the years since the invasion, smart devices have put the world in the palms of our hands, accelerating the speed at which different perspectives of war emerge. These devices liberate people from their immediate expectations but also create new contexts that disorientate even as they create new ways in which meaning is generated from online experiences. This is particularly noticeable in relation to social media, where, as context collapses, so does a sense of time (Brandtzaeg and Lüders 2018). The 'broken context' that this process creates is an effective way of describing the current situation in which war has expanded and meaning has become confused. Consequently, there has been a shift from a liberating sense of a free-for-all internet in the early 2000s to one in which every message and person is suspect in the 2010s and 2020s. Thus, Radical War has a paradoxical character: it is out in the open and part of the everyday and yet rendered invisible through its hyper-mediation.

For instance, Sara Wachter-Boettcher explains,

> It's not that technology broke my trust – at least not at first. But it broke my context: I don't know where I am. I don't know whether I'm at work or at play, whether I'm watching the news or chatting with a friend. This used to feel freeing: I didn't have to choose. I could simply *exist*, floating in a mix-and-match universe of my own design. But left unchecked for so long – by shortsighted tech companies, and by my own petty desires – that lack of context bred something sinister: a place where everyone's motives are suspect. (Wachter-Boettcher 2018, p. 39)

An example is the Russian involvement in the waging of an overt yet undeclared war against Ukraine in the second half of the 2010s. Having invaded the Crimean Peninsula in 2014 by *coup de main*, Russia went on to sponsor attacks against the Donetsk and Luhansk regions of western Ukraine. On the face of it, the conflict in Donetsk and Luhansk has the appearance of a traditional 'hot' war of trenches, artillery and tanks. Combatants define frontlines and fight across no man's land. In this context, smartphones become an attack vector, enabling separatist hackers to target Ukrainian personnel with malware, making it possible to geolocate and attack artillery units (Arquilla 2021). At the same time, a more covert war of trolls, smartphone viruses, cyber-attacks and social media propaganda attempts to shape the news agenda and reframe perception (see Hoskins and O'Loughlin 2015; Patrikarakos 2017; Singer 2018; Hoskins and Shchelin 2018; Pomerantsev 2019).

Such activity does not emerge spontaneously but demands organisation and direction. Thus, the pro-Kremlin Internet Research Agency in St Petersburg hires hundreds of cyber-workers for a 'Troll Farm' to produce online stories, videos, photos, memes, comments and contributions that promote Russian interests in Ukraine. These farms and cyber-operatives create and disseminate explicit 'fake news', but they also create fake profiles ('sock puppets') as part of a campaign to propagate division and stoke fear; all as part of a process they hope will lead to uncertainty, confusion and inaction both on the frontlines and among civil society beyond.

On the other hand, this implies that military practice has evolved to take advantage of digitalisation and the internet. And yet Russia's (un)declared overt/covert war against Ukraine has also been usefully characterised by Metahaven (2015) as a kind of 'black transparency'. A hallucinatory machinery of fantasy, fiction, antagonism and glamour described by Peter Pomerantsev as a permanent spectacle where 'nothing is true and everything is possible' (Pomerantsev 2015, p. 6). As an information strategy, the muddying of war's finitude creates uncertainty and clouds processes of sense-making. This in turn gives oxygen to Radical War, feeding the crisis of representation that the strategy aims to control, locking digital individuals into social media prisms.

Working in this broken context has left traditional broadcast news media journalists flatfooted, caught in a moral panic over fake news, indulging in nostalgia for how journalism used to be (Farkas and Schou 2018). In the face of the latest outrage on Twitter or Facebook, people's trust and confidence in older modes of representation are in a state of flux. MSM have found it challenging to adapt to an information ecology that exploits social media for the purposes of spreading disinformation. And it is in this crisis of (mis) representation – in which broadcast-era and participatory journalism are structured and drive audience engagement in different ways – that we find war's legitimacy most open to question.

Military interventions in the 1990s were framed by a concern for preventing genocide and protecting human rights. In the twenty-first century, these motives gave way to more realist pre-occupations associated with the GWOT and Great Power competition (Blanken 2012). In the period since the start of the GWOT in 2001, the rapidity and spread of data and information about war ought to herald a democratisation of perception in which multiple views of the battlefield are available to anyone. However, a pervasive and continuous streaming of every act of war has not delivered an enlightened awareness and a new politics of intervention. Instead, Western publics have rejected overseas entanglements.[40] This cannot be attributed solely to the evolution of digital media. However, as social media has taken on its own narrative force, it is apparent that this has formed a vector for debating and developing counterarguments for engaging in overseas intervention. This has fuelled doubts about the effectiveness of intervention and discouraged Western publics from taking on the risks associated with the use of military force.

This is especially the case in relation to the wars in the Middle East during the 2010s. These have starkly revealed how digital technologies and the infrastructures of representation offer unprecedented access to the visage of war and at the same time also carry audience-participants to the frontline. The sum total of these digital infrastructures facilitates the livestreaming of battle and in the process creates an accumulative and searchable archive of information on an unparalleled scale. As a consequence, the digital archive is one of the principal new battlegrounds of Radical War; and in this respect, given the continuous uploading and sharing of events from an array of its participants, victims and bystanders, the 2011 Syrian Civil War is an exemplar. A simple YouTube search for 'Syrian war', for example, brings up a constantly changing number of videos, where the first five hits range from being uploaded in the last five hours to as long ago as two years; some have viewing figures in the millions.[41]

This may all sound like a fabulously democratic vision of war where transparency and knowledge imply that states and participants can be identified from the digital flows of information and, as a result, held to account. Far from producing transparency, however, Radical War in practice yields the exact opposite effect. The flow of data leads to ambiguity, opaqueness and cover-up, to multiple, fragmentary and discordant narratives about war. Unable to contain the monsters created out of social media, this process not only provides cover for traditional forms of war and genocide but also enables entirely new forms of warfare. In the process, it encourages individuals and dissident groups in the West to intervene and join in wars overseas. In this respect, Radical War is part of a wider ideology of digital openness (Hoskins 2018) but one that obfuscates the formation of a consensus reality of war and instead can be used to promote doubt and inaction.

This media of suffering and outrage has not translated concerns for human rights and the protection of civilians into military action. The Western shift from interventionist to sceptic is not the result of some kind of hyper-compassion fatigue (a hangover term from the broadcast era) of watching children die in Syria every day for years and years. Rather, it is disinformation that is the convenient excuse for inaction in Yemen or in Syria. As Pomerantsev (2019) acutely observes, to what extent can we claim that '[w]e didn't do anything because we were confused by a bot farm'? This is the lesson that the international community have had to learn: that the hastily proclaimed democratising wave of the so-called 'Arab Spring' in the early 2010s has not furthered an enlightened politics produced out of an embrace with Silicon Valley's globalising technologists. On the contrary, the Arab Spring now appears to be the high point in the spread of Western values.

When twenty-five hospitals in Idlib, Syria, were bombed by the Syrian government and Russian forces over the course of one month in 2019, there surely must be some reflection on the relationship between the instant availability of billions of images of human suffering and death in the continuous and connective digital glare of social media, and their effect on global actors charged with the preservation of civilian life.[42] It is easy to conclude that any such relationship does not in fact exist and is born from an unrealistic or mythical expectation of some kind of functional journalism. A key consequence of Radical War then is the utter paralysis of humanitarianism.

In these circumstances, it is not surprising that amid the twenty-first-century's data storm, the (in)visibility of civilian deaths (particularly in states subject to multiple international military actions such as Iraq, Syria, Libya and Yemen) is central to the battle over legitimacy in Radical War. Investigating the mass of open-source and other data available on civilian casualties, war-related migration, refugee status and asylum-seeking is neither a priority for those responsible for civilian deaths and injuries, nor for many of the Western news media editors charged with overseeing reporting on these war zones. For the last thing that MSM audiences want to engage with is the reality of their governments' actions in producing death in both distant and at the same time domestic parts of the world. Equally, '[i]n the absence of a relevant editorial mandate, U.S. media professionals describe civilian harm reporting as siloed, fragmented, and as largely self-directed by individual journalists'.[43] This in turn reflects the US military's ambivalence to tracking civilian harm, which, as one military journalist for a major US newspaper observes, points to how, '[o]n the front end, they put in real effort to prevent civilian harm. What they are not interested in is grading their homework on the backend.'[44] Paying lip service to how civilian harm is represented thus tells us something about the evolution of the military–industrial–media complex in the twenty-first century (Der Derian 2009).

It is possible to critically assess the way military operations unfold. However, as Airwars – a not-for-profit NGO that tracks, assesses and archives international military action in relation to civilian harm – notes, MSM's willingness to critically engage with the 'backend' of military operations is noticeable by its absence. This contrasts with the emergence of a community of new, open-source intelligence organisations such as Bellingcat, the Syrian Archive, Action on Armed Violence, Forensic Architecture and indeed Airwars itself. In these new organisations, it is possible to see the establishment of a new normal in relation to investigative journalism about war. Online, and geared up to process the interrelationship between events and the saturated data environment, these organisations are part of and reflect what we call the new war ecology.

For organisations like Airwars, the challenge is to retain the integrity of their findings in an environment in which datafication has imploded the traditionally discernible separation between actors (journalists, citizens, soldiers, states, militaries), representations and acts of war. Engaged in a process of identifying the effects of the air campaign in Syria, Airwars aim to log and investigate every case of civilian harm caused by ordnance dropped from aircraft. Drawing on methodologies that have parallels with defence intelligence analysis, Airwars have effectively developed their own open-source process for conducting battle damage assessment. The result is proving to be particularly corrosive for armed forces who now have to manage their own narrative in relation to that put out by Airwars.[45]

The use of online data to help develop their assessments inevitably opens Airwars up to deliberate disinformation attack. Such attacks involve intentionally falsifying the digital record to confuse or undermine the integrity of their approach to open-source intelligence. This in turn forms part of a wider counter-narrative designed to suppress public concern for the Syrian Civil War by implying that reports on the conflict contain falsehoods. The only response to this is for Airwars to rigorously apply their battle damage methodology and be transparent about their sources.

Airwars' investigations have highlighted the effects of the air campaign in terms of civilian harm. The upshot of this is that it has forced armed forces to provide a public account of their use of air-dropped ordnance with a view to explaining why civilians have been targeted. As a result, the US-led coalition have had to admit to unintentionally killing 1,321 civilians by airstrike in Iraq and Syria. This still falls considerably short of the 8,135 to 13,032 civilian casualties that Airwars estimate from locally reported sources based on 1,451 separate alleged incidents and is significantly short of the total reported by civilians themselves, where the number of deaths from airstrikes stands at between 19,080 and 29,426.[46]

But Airwars have not just questioned the legitimacy and legality of the coalition airstrikes. Their investigations also call into doubt the battle damage assessment methodology in use by US-led forces. In this way, the datafication of the battlefield has not only served to delegitimise airpower but has also opened up questions about how the armed forces try to control the public understanding of military effectiveness. Thus, the digital revolution goes beyond globalisation and process re-engineering or, in more media terms, reframing spectacle and narrative. It also challenges how we come to think about the efficacy of military power in the twenty-first century. Military claims are then levelled to the status of all claims made in the emergent post-trust environment, subject to continual challenge and counterchallenge, legitimately by citizens on the ground and by NGOs, for instance, but also suffocated by rampant disinformation as part of many people's daily media diet. And it is here where Radical War thrives, exploiting data overload and obfuscation to undermine the very declaration of goals and the demonstration of the realisation of those goals, once the basis of credible military threats.

The nature of this challenge is such that it cannot simply be resolved by tightening the way armed force is applied. For the steady accretion of a more detailed body of military doctrine is insufficient to resolve a challenge produced by an explosion of data. Indeed, as the wars in Iraq, Afghanistan and in Syria and Ukraine have shown, military doctrine's focus on coordinating military effects to produce tactical results – destroying bridges to cut supply lines or erecting fortifications to protect lines of communication – is but one aspect of contemporary war.

As recent wars have revealed, of even more significance is the way that war is presented and framed as media spectacle, designed to catch the eye, motivate participants and attract recruits. IS has been particularly adept at this, developing sophisticated and well-produced social media messages designed explicitly to attract foreign fighters to join the ranks of the vanguard who were building a new state out of Iraq and Syria (Ingram, Whiteside and Winter 2020). In this respect, the performativity of war and its fetishisation of explosions, martial dress and hardware as they are shaped through new and digital fields of perception is equally, if not more important. Inevitably, armed forces have sought to manage the application of force even as they try to take control of the narrative. However, Radical War stretches beyond the power of a military doctrine whose focus lies in information war.

This is most obvious in the US, where the routinisation of war in relation to its everyday and constant appearance in the media has paradoxically rendered everywhere war invisible. Even as social media companies have lost control over their data, as John Louis Lucaites and Jon Simons argue, it is war's 'normalisation that has rendered it simultaneously hyper-visual and unnoticed' (Lucaites and Simons 2017, p. 3). It is not only that digital media and an overload of data have distorted perception, but that this distortion operates along a spectrum from being hidden to being in plain sight. It is the latter – the saturated visibility of war and its consequences – that no longer functions in ways still expected of it. This misplaced faith in the power of the image of suffering offers the greatest cover for the continuation of Radical War, and it is in this space where the military cannot take control of the logistics of perception.

The perceptions of war have thus both been obscured and captured by this data explosion. Sometimes the goal is to purposefully obscure intentions, to hide the reality of war in parts of the world that the state would prefer us not to know about – like the British government hiding its provision of weapons to Yemeni forces by providing them to Saudi Arabia while denying their end use in the war in Yemen.[47] Sometimes it is just that war is obscured through overexposure, blunting the sense-making capacities of even the most seemingly hyper-informed of digital citizens. Sometimes, like the British Royal Air Force's dubious admission to only having killed one civilian in their air campaign in Syria, it is obscured by military bureaucracies who choose not to look for data so they can avoid challenging questions.[48] And sometimes social media companies themselves are captured by war as they seek to generate user engagement through the surfeit of information about war.

The way in which contemporary information infrastructures are organised implies the permanence of digital data. However, as Airwars calculate, the data about particular military events in

Syria remains stable on the internet for only about a year.[49] After that, data decays: video files get deleted and hyperlinks break. Yet, the intentionality of archival accumulation is further complicated by the emergent array of uses to which digital traces can be put. And this process of decay in effect accentuates certain voices and reframes debates that shape the construction of future narratives. Consequently, digital material facilitates the open-source reporting of war. But it does so on the basis of a fragile and decaying information ecology that is itself derived from an intentional desire to store data and is a product of the accidental process by which data is accumulated and breaks. These features are not just the result of individuals and military institutions losing control of their own data but are baked into the structures of the internet. This in turn lends online narratives to misrepresentation and disinformation.

Thus, the 'accidental archive' of the internet (Moss and Thomas 2018) and social media has enabled new NGO-style actors to probe and preserve war crimes, and to challenge 'black transparency' (Metahaven 2015). But the extent to which even these actors' forensic efforts offers a stable and complete account of events is doubtful. Surveillance, digital decay and the accidental archive, especially in the face of those working to propagate disinformation, cloud causal relationships and lead us to question the legitimacy of armed force, even as it draws us toward conspiracy theory.

In this context, what kinds of witnessing and evidence of war crimes might be useable by international lawyers? When stoked by propagandists looking to exploit the underlying structures of the internet, it is easy to see how twenty-first-century disinformation not only confuses and complicates reality but at a more fundamental level also leads many to question the relationship between military activity and political effect. Armed forces may produce one form of evidence that justifies their actions. Nevertheless, their narratives must run the gauntlet of social media. Here the profusion of voices and imagery quickly leads the online world to form opinions and draw its own conclusions, shaping the way politicians respond or try to lead discussion. In this respect, as President Trump's November 2019 pardoning of US Navy SEAL Edward Gallagher reminds us, even democratically elected leaders are willing to overlook illegal transgression if it helps to drive Twitter debates that alienate political adversaries and solidify voter support.[50]

Putting aside Trump's effort to take political advantage, the Gallagher case also tells us something about military culture and the digital archive. Gallagher was a Special Operations Forces chief convicted of illegally posing for a photograph with a dead Iraqi prisoner following the Battle of Mosul in May 2017. Given the way the photographs were being digitally shared, Gallagher clearly had no fear that his immediate circle would out him as someone who would break the law.[51] However, at the point where these images became obvious to audiences beyond his inner circle, Gallagher's SEAL colleagues testified against him. In this case, the assumption that armed forces should operate within the constraints of the law was eventually reinforced by the reproducibility of digital media.

The digital archives of war are, therefore, unevenly distributed and not equally available to everyone. Both the Gallagher case and the example of the Syrian Civil War reveal several hidden challenges in the way that data about these wars are stored, indexed, archived and retrieved for later use. The digital archive's status in terms of its production, veracity, ownership and finitude, is very much in flux, adding to the ambiguity of Radical War (including its beginning and ending). The understanding and the legitimacy of warfare, especially in much of modern Western culture, is so centrally wrapped up in its commemoration and remembrance that it is impossible to ignore the astonishing scale, data decay and potential uses and abuses of the emergent digital archives of warfare.

These changes bring memory and history into a new conflict of convergence and competition. The memory and history of warfare are both utterly transformed by the explosion of data and the asynchronous reproduction of war in online spaces being turned on and turned over anew. In short, the past is being rewritten, pulled down and ripped apart, not in what were once thought of as collective ways – as societies have always recast their pasts in light of present needs – but instead in ways driven by polarisation, division, exclusion and (de)globalisation. In this way, we can speak of a 'radicalisation of memory',[52] the past being invoked, repressed or remade as a key weapon of war, and central to how both past and ongoing wars are legitimised or delegitimised anew.

The archive has long been seen as the supreme medium, as the external and institutional basis for the remembering and forgetting of societies at different stages of development across history; as an ultimate storage medium and metaphor of memory; as the arbiter of truth and the means by which to hold government to account. But today the archive itself is networked, connected, mobile and can be carried in your hand. So, whereas the traditional archive was often seen as history's primary collaborator in marking a buffer, in drawing a line, in the creation of a repository of what went before, the digital archive collapses the temporal parameters of war. The idea of 'the age of perpetual war' (Kennedy 2015, p. 163), as we see it, is not only the perception of a stream of concurrent wars, but rather that war online is never at some kind of archival endpoint that would enable a hierarchy of understanding and impose a chronological order. If military operations never stop, then when might armed forces take a moment to take stock and learn from what has happened? Instead, the digital archive is always being recycled, reactivated, reused, renegotiated and reimagined and in real time. In this sense, Radical War is also archival war.

This is a telling problem for the armed forces, for not only does the archive inform the military's approach to learning but it also forms part of a process of maintaining accountable government. Thus, control over the 'paperwork' produced by the state constitutes an elementary building block by which the executive maintains legitimate authority over the levers of power. As digitalisation rips through the centres of bureaucracy, refashioning processes of data collection and storage, even as it hollows out more stable approaches to analogue archiving, we find that armed forces have lost control of the data they produce. This exposes a gap between what governments have said and what armed forces do.[53] The abundance and flux of data beyond government control fills this gap and can be exploited by those seeking to delegitimise martial activity and create doubt about the effectiveness of military power.

Control

From the perspective of the technologically developed world, the overload of data and information normalises war. This distorts and becomes part of the everyday digital fabric. This might lead us to conclude that governments, armed forces or the companies that own these new information infrastructures can control what now forms a core part of our everyday experience. That this is not the case is most easily seen when examining the dangerous repercussions that emerge when Western social media platforms are dropped into the media ecologies of different parts of the world.

In 2018, for example, Facebook admitted its role in spreading hate and encouraging genocide in Myanmar in 2016–17.[54] With violence spiralling, 725,000 Rohingya Muslims in Rakhine state fled after the community was subject to indiscriminate killings, the destruction of their homes, gang rape and other acts of violence. Facebook independently selected the consultants BSR to investigate whether the social media platform had contributed to this outbreak of political violence. BSR concluded that Facebook had polarising effects on Myanmar's population that undermined human rights.[55] This further distorted sectarian perceptions, stigmatised the Rohingya and helped incite the genocide.

For most of Myanmar's internet-connected citizens, Facebook was synonymous with the internet. In 2009, less than 1 per cent of Myanmar's 50 million people had a smartphone or internet access at home. As the country moved from being a closed society to one that allowed more freedom of expression and engagement with people, goods and services from overseas, access to the internet steadily increased. By 2018, 34 per cent of the country had access to the internet and 73 per cent of the population were using mobile phones. By January 2018, Facebook users in Myanmar numbered 20 million.[56] Eventually forming part of an extended battlefield, the political violence between a minority Muslim community and the majority Buddhist community of 51 million was accelerated by Facebook.

The Myanmar security forces' persecution and genocide of the Rohingya, a stateless Muslim minority, had been going on for some time. From the mid-2010s, however, this hate campaign found its way on to Facebook, where the Muslim community was regularly portrayed as an existential threat.[57] While the Myanmar government and military quickly took advantage of the penetration of Facebook across the wider population, weaponising narratives to stir up hatred, the company itself appeared unable or unwilling to intervene with any urgency or effect. Consequently, the most spectacularly successful Western-created and owned communication tool of this century had been dropped blindly into a country with little consideration of how it might be used to weaponise ethnic, religious and political tensions. The results were catastrophic, leading the journalist Aung Naing Soe to claim that if Facebook didn't see the frenzy of hate speech on their own services it was because they didn't understand the ethnic and social politics in Myanmar and thus 'didn't know where to look'.[58] Even in 2021, Facebook still cannot properly 'police content in multiple languages around the world' because the safety systems the company has in place are 'developed primarily for American English' speakers.[59]

In 2018, the United Nations Human Rights Council (UNHRC) International Fact-Finding Mission on the genocide in Myanmar reported: 'In a context of low digital and social media literacy, the Government's use of Facebook for official announcements and sharing of information further contributes to users' perception of Facebook as a reliable source of information.'[60] Paradoxically, while much of the West experienced a collapse in trust in mainstream and social media forms, in Myanmar the general public were prepared to invest trust in a media that would eventually prove to be lethal. Indeed, localised trust in Facebook was found to be a key factor in the spread of hate speech against Muslims (and the Rohingya in particular), helping to fuel human rights violations against them by mostly military and security forces in Myanmar.[61] Consequently, the UNHRC Mission report on the genocide in Myanmar recommended that '[b]efore entering any new market, particularly those with volatile ethnic, religious or other social tensions, Facebook and other social media platforms, including messenger systems, should conduct in-depth human rights impact assessments for their products, policies and operations, based on the national context'.[62] This is because, as Amanda Taub and Max Fisher explain, 'its algorithm unintentionally privileges negativity, the greatest rush comes by attacking outsiders: the other sports team. The other political party. The ethnic minority.'[63]

The UNHRC Mission report on the genocide in Myanmar reveals the difficulties in arriving at a comprehensive vision of the effects of any one digital media ecology. This is in part a function of the resources required to fully grasp the sheer scale and amplification associated with the reposting of hate messages. At the same time, social media platforms offer users the option to report content and moderate posts, a function that in turn plays a role in self-censoring and muting voices that would otherwise be heard. This dialectic between reporting content and muting voices contributes to the proliferation of social media platforms as users spill out from one site to another, trying to find an online space that tolerates their views. And this further complicates the media ecology, as hate speech is reproduced in cross-platform messaging and communication services, which, for those countries without sophisticated electronic surveillance capabilities, are inaccessible to surveillance and detailed investigation.

In these circumstances, governments are forced into blocking parts of the internet in order to prevent hate speech and stop the proliferation of violence. However, not all governments have the same capacity to interfere in how the internet operates. Some countries like the US and UK have very advanced capabilities, allowing them to interfere in what a specific person might see and thus shape what they do. Other countries have to take more blunt approaches, stifling online information flows to make it easier for violence to continue without oversight and civic dissent. This is roundly condemned as an attack on free speech when it takes place in authoritarian states like the Assad regime in Syria. Unfortunately, the technical capacity to manage incitement to political violence is not evenly distributed. The result is that even democratic states like Sri Lanka engage in blocking the internet rather than countering specific webpages or the groups that build them.

Thus, for example, following anti-Muslim riots in March 2018, the Sri Lankan government temporarily blocked most social media.

Following a series of terrorist bombings a year later, the Sri Lankan authorities took the same action again, this time rendering Facebook, WhatsApp, Instagram, Snapchat, Twitter and Viber inaccessible. Both decisions were designed to disrupt the flow of misinformation about the attacks and prevent the distribution of hate speech that could further ignite violence. By 2019, however, this shutdown was seen by some Sri Lankans as demonstrating how US-owned digital platforms had created a monster they could no longer control. Indeed, as one reporter noted, '[t]he extraordinary step [by the Sri Lankan government to shut down several social media platforms] reflects growing global concern … about the capacity of American-owned networks to spin up violence'.[64] Indeed, by 2020, Facebook investigated its role in the violence and once again apologised for helping to spread hate.[65]

Even as Silicon Valley pushes globalising media ecologies on to parts of the world that do not share similar media trajectories to those found in the United States, Western traditional broadcast and MSM reproduce the very ambiguity that is the most effective weapon of contemporary war. Once seen as dependable for making the origins, consequences of and potential exits from war intelligible, MSM have been hacked and annexed in ways that reproduce the challenges faced by social media. In the face of this data tsunami, the Western MSM have shown themselves unable to deal with the challenge. Instead, they have largely thrown their hands up in despair at their inability to impose a pre-digital standard of verification in news-making on today's media ecologies. Consequently, as one foreign correspondent with a major cable news channel observed in an interview with Airwars,

> [w]hen the Syrian civil war revolution first kicked off there was a very strong reluctance to use any video that we could not identify ourselves. There was a ton of stuff being put up on YouTube – a lot of it vertical [aspect-ratio] video – a lot of it just cellphone video – crappy video … If we did use it, we were always very careful to say we can't identify where this video came from … But, in the last couple years of the war it was a free-for-all. If a video was put up on YouTube it was used.[66]

Unable to use the information that comes from citizen journalists or using the material but heavily caveating the sources, MSM have created an opening for citizen journalists to fill an informational void that traditional editorial policies cannot address.

Such informational voids and gaps in meaning betray how government, the military and MSM are no longer in control of the media narrative. This became abundantly evident during the Iraq War. Described as the first YouTube war, the extensive use of digital cameras by US soldiers started to reveal the visceral nature of combat operations in Iraq (Andén-Papadopoulos 2009). As these technologies became more sophisticated, helmet-cam footage made it on to you YouTube (McSorley 2012). Of course, soldiers filming themselves is not new. However, following YouTube's launch in 2005, service personnel could also upload what they'd recorded for the world to see. This challenged Western MSM's representation of the war for audiences at home (Silcock, Schwalbe and Keith 2008). At the same time, it also helped Iraqi militants to understand how to position Improvised Explosive Devices (IEDs) and site ambushes so that they could be filmed in such a way as to gain maximum media spectacle.

Once the smartphone appeared on the market, the relationship between battle and its participants iterated once again. This time, the November 2008 terror attacks in Mumbai could be caricatured as Twitter coming of age (Ibrahim 2009). Taking place over sixty hours and involving more than twenty terrorists, the attacks killed around 170 people and wounded as many as 300 in a coordinated shooting and bombing spree across thirteen Mumbai locations. All the attacked locations were soft targets. At no point did the terrorists attempt to overcome guards. Instead, they organised themselves into teams and used their AK-56 assault rifles, pre-made IEDs and hand grenades to herd groups of people into killing zones, eventually taking thirteen hostages, five of whom were murdered.[67]

In terms of social media, it is estimated that the Mumbai attacks generated around seventy tweets every five seconds (Ibrahim 2009). Smartphone users disseminated images, videos and sound recordings while they were under attack. This in turn made it possible to get a real-time picture of what was happening across the city. A few hours into the attack, a Google map had been created that revealed the locations of the main incidents, and a Wikipedia page was created that provided background information. At the same time, Indian television stations amplified videos that had been posted via Twitter, ensuring that the trauma crossed over between new media and MSM. Equipped with Blackberry smartphones and a satellite phone, the terrorists coordinated their activities with each other and through a command centre in Pakistan (Kilcullen 2013) that was monitoring the various media feeds. This helped the terrorists find and kill groups of people who were trapped in various places (Oh, Agrawal and Rao 2011).

Following the Mumbai attacks, the style of attacks has changed again. In 2015, Paris was subject to a series of coordinated attacks by Islamist terrorists. Here three groups of men launched six attacks that involved suicide bombings and shootings across four locations. The attacks started at Stade de France, France's national sports stadium, took in a number of apparently random shootings and culminated in the Bataclan theatre massacre where ninety people were killed. The attackers ultimately killed around 130 people and wounded over 400 others.[68]

Like the Mumbai attacks, the Paris attacks aimed at maximum carnage using explosives and assault rifles. Instead of trying to find symbolic targets, the terrorists were directed to '[h]it everyone and everything'.[69] The goal was to indiscriminately kill rather than to make a complex political point. The act of killing would be political enough. The difference in the attacks was that the terrorists recognised that French intelligence agencies were better able to track phone calls and internet usage. Accordingly, they either used disposable phones with maps of the attack locations pre-loaded on to them or the phones of those people they'd attacked or even a specially encrypted laptop that one team member carried.[70]

More recently still, there have been attacks where vehicles have been driven into people walking along the Marseille beach front, killing eighty-six and injuring over 450. Bombings at Brussels airport and Maalbeek metro killed thirty-two and injured 300 in 2016. The bombing of Manchester Arena killed twenty-two people in 2017, and countless other attacks across Europe have also been designed to maximise death and spread fear among the public. All these attacks have caught the security services and the police by surprise. All of the attacks have been enabled by smart devices. All of these attacks have sought to achieve political effects by maximising the media spectacle.

In Radical War, the changing terrain of battle reflects the changed relationship between its participants and the connected technologies that have re-worked societies across the world. Coordinating terror activity by smart device might not be particularly surprising in and of itself. What is more interesting to reflect on, however, is the possibility that the attacks demonstrate an awareness that war has moved away from simply attempting to capture and dominate terrain. Instead, these terror attacks signify that the conduct of war is now more focused on identifying the social terrain that people operate within. This is much more transparent than it has ever been. Now people make explicit the networks they belong to through, for example, the Twitter users that they follow. This new terrain may well include a physical location, but it also points at the digital individuals and their social networks. The terrorist is seeking to operationalise these online networks to advance their political agendas. In both situations, social media have accelerated the speed at which gruesome images of war are posted online. The attackers have not had to broadcast themselves. They have instead been able to rely on others broadcasting events for them. This has in turn increased the value of escalating violence for even more gruesome effect. All of this implies that future battlespaces will proliferate as attackers seek to gain political advantage by grabbing attention in increasingly horrific ways.

The counter to this position is the way that a fragmentary global civil war has normalised violence such that

> [w]ar is becoming normal: the stock exchange no longer reacts to massacres, as its main concern is the looming stagnation of the world economy … [and after every attack] … whether by Islamists or white supremacists, by random murders or by well-trained fundamentalist killers, Americans run to buy more weapons.[71]

Here the suggestion is that society has found itself stuck in a violent loop from which there is no obvious way out, where it is unclear how representation and reality relate to each other. In this new dystopian world, what was once viewed as conspiracy is now considered to be the new normal. Indeed, conspiracy has been drawn into the day-to-day functioning of the political process and is now a core driver of political change. Maintaining control in a context in which white is black and black is white demands an engagement with cybernetics that Western governments have yet to manifest in any meaningful way. In this context, Radical War suggests that the new war ecology is beyond control.

Paul Virilio offers us a measure of how far-reaching Radical War is in its representation of contemporary conflict and violence. In 1989, Virilio wrote that 'the history of battle is primarily the history of radically changing fields of perception' (1989, p. 7), such that '[t]here is no war … without representation, no sophisticated weaponry without psychological mystification' (1989, p. 6). Radical War thrives precisely because the fundamental relationship between perception, knowledge and action has changed. These changes are rooted in the way that politics, society and the economy have transformed since the turn of the twenty-first century, heralding information infrastructures that create digital prisms that are hard to escape from. In this new 'post-trust' context (Happer and Hoskins 2022), it is difficult to discriminate between information and disinformation.

This reality has been reproduced on and through the social media technologies, networks and companies that billions of individuals have become so reliant upon. These new information infrastructures form the basic pillars of social life, of identity and work. Yet people have little understanding of the coding and the algorithms that make these platforms work and that enable the trolls, deep fakes and hacks that manipulate attention. Consequently, how we perceive the world is in a state of dystopian crisis. Reality and its representation appear to have collapsed, inverting into each other, challenging how meaning is generated and the effects of war are to be understood. In what follows, we investigate how we got into this position and what it implies for war and society.

2

UNDERSTANDING THE NEW WAR ECOLOGY

War in the twenty-first century is participative. It is war without bystanders. By this, we mean the process of networking individuals and their digital devices has made them both part of and subject to warfare. At one and the same time, people can record war and unwittingly transmit datapoints that are useful to those engaged in generating targets for the battlefield. Web 2.0, smart devices and the IOT create new 'architectures of participation', enabling a wide range of actors – militaries, states, journalists, NGOs, citizens, victims – to have their say and participate in warfare in an immediate and ongoing fashion (Merrin 2018). But this very act of participation collapses the boundary between those who observe war and those who engage in it, lulling actors into a false sense of being active, of making a difference, creating shaky expectations that information translates into both knowledge and action. In their continuous production and regurgitation of data, these participant-combatants make new kinds and new scales of war possible. As a result, war's traditional parameters, where wars were fought over regimes, religion, territory and the economy, have been reconfigured around the digital individual. At the same time, our daily interactions on and with digital devices and networks have produced an excess of data, presenting an opportunity for those wanting to exploit this for the purposes of surveillance, policing and warfare.[1]

Although we can still see echoes of the twentieth-century media ecology in which various elites could command the attention of audiences through broadcast media, in the twenty-first century the digital individual occupies an environment where data remakes and displaces the centrality of the MSM. Digital individuals now have the technology to both produce and consume material that they and others have created. This has led people to create their own media channels, hosted on web platforms like YouTube. This heralds a shift from a long-standing set of power relations framed in terms of media-attention-control to what we view in terms of data-attention-control. This has displaced the position of broadcaster and replaced them with the user, where the user is now central in creating and transmitting narratives and its associated metadata. These changes signal the emergence of what we call the new war ecology.

One fruitful way to make sense of these changes is to consider Benjamin H. Bratton's meditation on how planetary-scale computation has remoulded geopolitical realities (Bratton 2016). Bratton proposes a model he calls 'the stack', which not only offers an 'alternative geometry of political geography' but also demands that we 'map a new normal' (Bratton 2016, p. 4). The new normal, then, of the 'new war ecology' is the rapid emergence of a hyperconnected environment in which datafication implodes the traditionally discernible separation between actors, representations and acts of war. Consequently, the hyperconnected is now both immediate and asynchronous, compressing time into the permanent now that is best represented in the rolling social media feeds that keep users locked in the moment and that have supplanted MSM.

In this chapter, we map the emergence of this new normal, a dynamic paradigm or 'ecology' of war, to begin to appreciate how to apprehend Radical War. We adapt the term ecology from a long tradition of work on 'media ecologies' (McLuhan 1964; Postman 1970; Fuller 2007) and draw on Andrew Hoskins and John Tulloch, who define media ecology as

> the media imaginary (how and why media envision the world within a particular period or paradigm and its consequences) *and* our imaginary of the media of the day (how media are made visible

> or otherwise in that process of making the world intelligible), in which some ecologies are perceived as inherently more 'risky' than others, by news publics, journalists, policymakers, and scholars. (2016, p. 8)

Our development of a 'new war ecology' takes these influences but goes further. We use the term to offer a new way of exploring how war is utterly saturated by data and fought and experienced within an 'information infrastructure' that mixes the human and the non-human (Bowker and Star 2000). The new normal destabilises old modes of representation and trust, leaving knowledge about and understanding of war in a state of flux. The tsunami of digital narratives has confused audiences and the MSM as to which version of events to believe. This has made war seem much more fluid, co-present, persistent, immediate, and in the volume of contradictory material that is now available, resistant to intelligibility. With highly political and greatly contested results, it has also led to the emergence of a new data economy, profiting from the collecting, storing, aggregating, hacking and the buying and selling of personal data. This is a world of ubiquitous surveillance in which most of what we do 'has at least the capacity to be observed, recorded, analyzed, and stored in a databank' (Cheney-Lippold 2017, p. 4).

Just as importantly, by enabling participative war (see Appendix), this chapter sets out how technologists in Silicon Valley have disrupted traditional twentieth-century modes of civil–military relations. This previously depended on a clear distinction between those in uniform, civil society more generally and the politicians to whom the military reported. By removing the bystander from war, however, the traditional models of civil–military relations are having to contend with technologists redrawing relationships between armed forces and society. When everyone participates, how do you distinguish between civilian and combatant? This change more than any other is having a radical effect on how we must think about war in the twenty-first century.

In this chapter, then, we seek to identify the main features of the new war ecology and explain how it contrasts with an older twentieth-century way of conceptualising war and media. This will establish the main analytical framework for examining how our fields of perception have become thoroughly saturated by processes of digitalisation and datafication. The net result disorders older modes of representation and demands that we think again about the mutability of knowledge in relation to the legitimacy of war as it is depicted in history and memory. At the same time, by showing how older models of civil–military relations are being disrupted by patterns of participative war, we establish the basis for how we investigate data, attention and control in the rest of the book.

The Old War Ecology

Scholars of media and representation have sought to highlight the importance of the mediatisation of contemporary warfare (Cottle 2006; Hoskins and O'Loughlin 2010; Maltby 2012; Patrikarakos 2017; Singer 2018; Merrin 2018). But for many strategists and scholars working on the significance of military power, the media tend to be thought of as ancillary to the military's main effort of securing a government's war aims. As adversaries have started to make greater use of the internet to spread their messages, however, armed forces have demonstrated a greater concern for the information environment and the evolving relationship between war and its representation in the media. As a consequence, the identification and labelling of new models of conflict have multiplied. Mostly, these analyses work within a twentieth-century paradigm of war and media, where the predominant model is framed in terms of broadcasting media rather than individual participation. As a result, scholars tend to look for continuities with existing interpretations of war, explaining away what is new by drawing parallels with the past in the hope that the armed forces can reorganise themselves to more effectively manage risk, chance and uncertainty.

There is of course a history of theories of so-called 'new war'. First published in 1999, Mary Kaldor's stands out in the field (Kaldor 1999, 2007, 2012, 2013). She writes that '[i]t is the logic of persistence and spread that I have come to understand as the key difference with old wars' (Kaldor 2013). Although scholars have found much to criticise in the concept of New Wars, Radical War is not an update to Kaldor's thesis but rather a fundamental break with it. This is partly because an incredible new battlespace of social media of unprecedented complexity, scale, persistence and spread has emerged in ways that have certainly had effects on how wars are fought. More than this, and as we will explain below, processes of digitalisation are fundamentally disrupting twentieth-century models of civil–military relations.

The New Wars thesis fails to capture the dynamics of the new war ecology as a contradictory space that breeds Radical War. For the most part, then, books about war only tangentially consider war and its representation. This becomes obvious when looking at the switch from explaining war in terms of state versus state models of warfare – models that emphasised the way in which technology has structured and made intelligible the chaos of battle (Bousquet 2008; Lindsay 2020) – to those that have tried to make sense of global insurgency, terrorism and political Islam (Kilcullen 2009; Devji 2005, 2009). While the literature on insurgency, terrorism and the challenges posed by political Islam has primarily been framed by 9/11 and the GWOT, its origins lie with the 'new and old wars' debate (Fukuyama 1992; Huntington 1996; Kaldor 1999; Shaw 2003; Münkler 2005) and the US military's desire to lead a Revolution in Military Affairs. The idea of a Revolution in Military Affairs has a long trajectory reaching way back to the mid-twentieth century, but the goal has always been to ensure that America's military capability was ahead of the rest. The result has been to pursue technologies that took advantage of, for example, precision, stealth, digital connectivity, networks and the revolution in intelligence, surveillance and reconnaissance (Arquilla and Ronfeldt 1993; Krepinevich 1994; Hundley 1999; Rasmussen 2001; Lonsdale 2003; Kagan 2006; Coker 2012).

After 9/11, as armed forces got to grips with counterinsurgency, the possibility of cyber-attack, influence operations and hypermedia, there has been a further iteration of literature away from discussions of counterinsurgency (Nagl 2005; Ucko 2009) towards strategic communications, propaganda of the deed and psychological warfare (Arquilla and Borer 2007; MacKinlay 2009; Bolt 2012; Freedman and Michaels 2013; Rid 2013; Briant 2015a, 2015b). Following the withdrawal of American and coalition forces from Iraq in 2011, this has given way to newer texts that have been critical of counterinsurgency (Porch 2013; Smith and Jones 2015) or sought to make sense of warfare in Syria, Crimea and Donetsk (Lister 2015; Hashim 2018; Fridman 2018).

Stimulated by the pace of change in how war is fought, we have subsequently seen a proliferation of labels for describing war and warfare. These 'new' paradigms reflect an uneasy recognition that the way wars are fought is changing more rapidly than our capacity to make sense of it. Thus, since the new and old wars debate of the early 2000s we now have fourth-generation approaches to warfare (Hammes 2004); global insurgency (Kilcullen 2009; MacKinlay 2009); irregular war (Rid and Hecker 2009); algorithmic war (Amoore 2009; Suchman 2020); proxy war (Hughes 2012; Mumford 2013); hybrid war (Hoffman 2007; Fridman 2018; Galeotti 2019); full-spectrum conflict (Jonsson and Seely 2015); ambiguous and non-linear war (Galeotti 2016); accelerated war (Kallberg 2018; Carr 2018; Horowitz 2019a); greyzone war (Echevarria 2016; Lohaus 2016; Jackson 2017; Wirtz 2017; Lupion 2018; Cormac and Aldrich 2018), shadow war (McFate 2019); full spectrum dominance (Ryan 2019); surrogate warfare (Krieg and Rickli 2019); liminal war (Kilcullen 2020); asymmetric killing (Renic 2020); information at war (Seib 2021); military AI (Johnson 2021); the warbot (Payne 2021); vicarious warfare (Waldman 2021); identity warfare (Jacobsen 2021); and thanks to the US Defense Advanced Research Projects Agency, mosaic warfare.[2]

Many of these models of war and warfare share a common root in that they try to explain or work out how to deal with a perceived collapse in the binary categories of war/peace, combatant/civilian, inside/outside. In this respect, they seek to explain war as a 'space of activity that is ethically ambiguous, with ill-defined outlines' and a 'complex internal structure' (Fuller and Goffey 2012, p. 11). In contrast to total war, these wars are rarely state versus state but now more typically are irregular and fought by proxy. The sinews of war reach out across maritime and trade routes (Khalili 2020) into the Middle East and occasionally bubble up as terror attacks within the metropolitan centres of Europe and the United States. At the same time, the logistics of war can be traced back and through peripheral geographies in Somalia, the Middle East, the Sahel, the Philippines and elsewhere. Adversaries are from overseas, and yet home-grown threats emerge from the connections that people make on the web. Wars involve technologies that were designed for conventional battle and at the same time are improvised, sometimes built from designs provided by foreign powers and yet can be made from commercially available products (Cronin 2020).[3]

The proliferation of books that seek to explain how war has evolved since 9/11 is revealing in a number of ways. Clausewitzian scholars, for instance, argue that the cottage industry in books on the changing character of war can be explained by a fundamental misunderstanding of war's nature. They argue that the binary categories of war and peace, combatant and civilian, inside and outside the state have not collapsed into each other (Stoker and Whiteside 2020). Rather, the fashionable labels that dominate contemporary thinking on war – hybrid, greyzone or liminal war – focus on how war is fought when they should be focused on the political objectives for which war is fought. Analysts are too focused on how big a war is, or what weapons are used. Instead, they should be thinking about why states go to war and what a war is for.

Inevitably, the focus on how wars are fought leads to a situation in which 'the tactical becomes the political, with the result that the point of war becomes war itself' (Stoker 2019). If war becomes an end in itself, then unsurprisingly war is perpetual. It is perpetual because analysts have no sense of what it would take to identify a political objective and then work out how to align the use of military power in order to achieve it. This is because they have no understanding of and indeed are not interested in defining the political objective of the war and thus cannot state what victory looks like. For Clausewitzians, then, an analysis of limited war that does not weigh up costs and benefits in relation to the political goal being pursued only leads to political and military incoherence.

According to strategic theorist Donald Stoker, the trajectory of poor thinking on limited war has a long history that starts in the early years of the Cold War and has continued to shape post-9/11 analysis. Framed by the possibility of a nuclear exchange between the United

States and the Soviet Union, the work of key American strategic theorists like Bernard Brodie, Robert Osgood and Thomas Schelling deliberately sought to restrain conflict in an effort to prevent war escalating to the point of atomic holocaust. These thinkers chose to frame their approach to limited war against the possibility of human annihilation to ensure that small wars did not turn into big wars. When looked at this way, it was important to restrict the political purpose of war in order to avoid confrontation. Instead, war was an act of signalling where belligerents had a common interest in which they sought to negotiate over acceptable war outcomes (Stoker 2019) to prevent escalation.

The question arises, then, of why Western leaders have failed to adopt military strategies that would result in achieving the political objective of the war. One reason for this is that political leaders have failed to properly identify appropriate aims for the wars that they choose to fight. Instead, insulated from the reality of their decisions by poor strategic theory repeated by academics and the foreign policy establishment, they hedge and hope to avoid telling the public why military power is being employed in the way that it is. The result is an unnecessary waste of resources and the needless deaths of civilians and soldiers. This implies that 'Western democracies have a deep-seated problem: their political and military leaders too often do not understand how to *think* about waging wars, and thus don't wage them effectively' (Stoker 2019). Worse is that they have failed to do this since at least the Korean War.

As a number of commentators have observed in relation to British strategy-making, the failure to think strategically is not limited to the United States but appears to be endemic. Thus, for example, Professor Hew Strachan writes that '[s]trategic theory has failed to provide the tools with which to examine the conflicts now being waged' (Strachan 2008, p. 51). At the same time, thinking about political ends rather than the military means has also eluded the British political-military elite. Consequently, Professors Paul Cornish and Andrew Dorman go back to first principles as they seek to explain the relationship between policy and military strategy as it relates to the Blair government's failure to match means and ends (Cornish and Dorman 2009). Patrick Porter goes further and observes that Britain has failed 'to craft strategy that reflects not just its aspirations, but its actual interests and capabilities', putting this down to an 'intellectual vacuum at the heart of British statecraft' (Porter 2010, p. 6).

In the wake of defeats in Iraq and Afghanistan, this has prompted calls for better strategy-making even as different parts of the political-military establishment have sought to deflect blame (Ford 2021). Here the focus is on writing a historical narrative that explains away the failures as a result of poor funding or a stab in the back. This has led some to conclude that a particular interpretation of the strategic mess that is Iraq and Afghanistan has formed into the narrative of the apologist. The claim being that politicians and civil servants were not up to the job of delivering victory in war (Dixon 2019).

In each of these cases, however, the prevailing academic response has been to try to shore up the Clausewitzian model and remind practitioners that they have allowed themselves to slip into shoddy thinking. What is not questioned is the possibility that the theory does not explain the practice. Instead, when politicians and the armed forces fail to realise their goals, they are criticised for not living up to the ideal. These criticisms are even more surprising given the amount of money that the United States and Britain spend on civil–military bureaucracies that are designed to support a Clausewitzian model of decision-making.

Unfortunately for those trying to buttress these features of the old war ecology, the challenge of integrating civil–military decision-making will only get harder as digitalisation upends traditional models of state bureaucracy. At the heart of the twentieth-century model of civil–military relations lies the central principle that civilian actors retain objective control over the military. In return, civilians do not interfere in military matters of professional space and judgement and the services stay out of politics (Huntington 1957). Although history shows that this is more the exception than the norm, if this relationship is harmonious, then there is the potential for states to develop coherent strategies. As processes of digitalisation sweep through government bureaucracy, however, there is even less certainty about how to sustain this model of decision-making. This is because technologists in Silicon Valley have already designed the platforms necessary for outsourcing key processes to the cloud, making it easier, for example, to maintain inventories, manage client relationships and sustain productivity for virtual teams.

Even more corrosively for existing models of government, companies like Facebook have designed a platform to do government more effectively than government itself. When the volume and varieties of data produced by these sorts of platforms are connected to AI, government will have the capacity to speed up decision-making (Harkness 2017). Consequently, problems and solutions can be identified more quickly, helping government organise responses across multiple communities and domains to resolve challenges and satisfy needs. Implementing this model of decision-making means changing the way government processes data. As Dominic Cummings, the former advisor to British Prime Minister Boris Johnson, says: 'Continuing with the Pentagon and the UK Home Office is "risky". Continuing with the leadership of the Met police and its management is risky. *Replacing them is safer*.'[4] According to this line of thinking, supplanting traditional patterns of bureaucracy with radically different operating models leads to better government. However, existing models of civil–military relations will have to adjust to these changes if they are to maintain Clausewitzian thinking in the new structures. While the bureaucratic challenge associated with harmonising decision-making might be surmountable, trying to sustain older forms of civil–military relations on the battlefield will be harder. This is a function of participative war, and it is to this subject that we return once we have explained how war and media is transitioning from the old to the new war ecology.

Mapping the Crisis of Representation

Identifying the transition from old to new war ecologies is not easy given the widespread availability, accessibility and pervasiveness of digital technologies, devices and media. The abundance of these media creates a profusion of different perspectives. Not only does this usher in a dizzying array of narratives, but it also points to a crisis of representation where consensus about war is in a permanent state of flux. One way of mapping this crisis of representation begins with acknowledging the importance of states in shaping narratives about war through official sources and influences over the course of time. Having traced the official narrative, it then becomes easier to show how different ways of seeing are reconfigured and enrolled to renew consensus or split off and become the means by which a shared understanding is undermined.

For the historian Jay Winter, the interaction between this official narrative and alternative and unofficial accounts of war represents a long-standing dialectic. This dialectic happens when conventional representations of war meet what the mainstream consensus might consider the transgressive. Mirroring Roland Barthes, Winter calls conventional images or received wisdom about war the 'studium' and that which disrupts or stands out the 'punctum' (Winter 2017, pp. 62–7). The interaction of the studium and punctum synthesises in ways that can have transformative results that change how war is understood. Winter, for example, notes that

> the conventional changes over time so that the images of dismembered bodies or of civilians as the targets of war, from being, as it were, exceptional have now become normal. If they are normalised ... then there is a shift from the imaginary of war as that which happens between soldiers to an imaginary of war that happens between everybody.[5]

Winter's work is crucial in revealing how counter-narratives reconfigure or become enrolled in otherwise consensus views about war and its relationship to the state and society. At the same time, this dialectical method offers a useful way to explore the crisis of representation as it has emerged in war and media since 9/11.

Retracing the impact of war on the collective consciousness of a society is ultimately both enabled and constrained by the mechanisms that render it visible, accessible and intelligible. So, for instance, the art-historian Ulrich Keller argues that the 'first media war in history' was the 1853–6 Crimean War. Although the transgressive narratives that might emerge from this innovation were relatively few, this moment still represented 'the first historical instance when modern institutions such as picture journalism, lithographic presses and metropolitan show business combined to create a war in their own image' (Keller 2002, p. ix).

By contrast with war and representation in the nineteenth and twentieth centuries, social media have exploited the global reach of the internet and expanded the battlespace into online and distant locations. This has given unique voice to the individual preferences and opinions of social media users. Although these are algorithmically charged, pushing extreme views, there is nonetheless an immediately accessible, always available archive of images and narratives about war. This is at a scale unimaginable from the perspective of a pre-digital era.

Societies have always recast their pasts in light of present needs (Lowenthal 2012), yet social media archives enable the past as well as the present to be invoked, repressed and regurgitated in contagious and polarising ways. This profusion reflects the studium and the punctum of war colliding and blurring narratives together. As a result, private and public, illicit and official, transgressive and conventional narratives all inhabit an unprecedented internet-enabled archive. Technology consequently facilitates and constrains the digital individual as they find ways to contribute towards and break the mainstream consensus over what constitutes knowledge and what ought to form part of the received historical record.

In these circumstances, if the new war ecology is in a constant state of flux, then attempting to 'fact-check' what is being regurgitated seems to be utterly absurd. Indeed, to try to reboot some kind of late twentieth-century MSM advocacy in the veracity of the spoken or written word or image labelled as 'news' is either misplaced nostalgia or an attempt to manage the punctum and studium of war for political purposes (Applebaum 2020). At least some are honest about a crisis in journalism.[6] For example, Jeff Jarvis believes that 'we are at the start of a long, slow revolution, akin to the start of the Gutenberg Age, as we enter a new and still-unknown age'.[7]

Thus, our model of a new war ecology, although concerned with the flux of data about war in the present, recognises how media, memory and history exist in a new nexus of unprecedented complexity and scale. This produces a relentless churning of the studium and punctum of war. In the next sections, we unpick this entanglement to reveal how the new war ecology has emerged out of a dialectical relationship between the pre-digital and the digital order of warfare. This will help to reveal a fundamental rupture in war's representation, perception and experience.

The Studium and Punctum of War From Vietnam To Pre-9/11

In the twenty years between 9/11 and 2021, there have been dramatic changes in how war is recorded, documented and experienced that have fundamentally reshaped what features of war are seen and attended to. To make sense of how these changes have led to the emergence of what we are calling the new war ecology, the next two sections adapt and make use of Winter's notions of studium and punctum in an effort to reveal how countervailing narratives in war and media interact and evolve over time. In particular, our goal is to show how Clausewitzian principles of war and the framing of war in the media have interacted in such a way as to produce a certain type of narrative about war within the West up until 9/11. We can then explore how 9/11 represents a break with older representations of war. Before 9/11, the studium could be characterised as an interrelationship between broadcast media, principally as framed by television news, and a conception of limited war that, however inaccurately, had been informed by Clausewitz. In the period after 9/11, the studium and the punctum of war have interacted in ever-accelerating cycles. This acceleration has created the conditions that have facilitated the emergence of the new war ecology.

The perception of war has been repeatedly reshaped by the ebb and flow of media and events, which is in turn shaped by an interplay between the media industry and the military's approach to war. As the technologies of recording, documenting and publishing have become accessible to everyone, the media industry's capacity to act in ways that reinforce conventional images or received wisdom about war has been fundamentally undermined. The process by which this has happened and the interplay between the studium and punctum of war explains why the emergence of the new war ecology continues to be masked by older perceptions of war.

Since the Crimean example above, war has become increasingly scripted, photographed, filmed and televised for the masses. The MSM production of war reached new mass domestic audiences with the US 'Living Room War' of Vietnam (Arlen 1966), and global real-time televisual spectacle with the 1991 Gulf War. News organisations, both helped and hindered by militaries and states, produced wars in ways that would suit their audiences. In the twenty-first century, this has been reversed. Now technology companies create web-platforms that allow participants to broadcast their stories about war. In effect, this is a *reversal* of the basic tenets of media production. Once audiences for war were the end point of the communication of news and information, at least in terms of a linear, reductive, yet influential 'broadcast-era' model of media studies (see Merrin 2014 for a critique). By contrast, audiences today are more like nodes in a network, part of a hyperconnected ecology of war, that constantly create and consume media but are not reliant on traditional broadcasters. The result is a relentless churning of differing opinions and images of war, such that consensus about what war is becomes much harder to construct and maintain.

Yet, despite these changes in the production and consumption of media in the twenty-first century, the contemporary history of war and media – in terms of its public and political orientation as well as in relation to whole fields of scholarly work on war – is utterly dominated by television news. This Western mode of perceiving war spills over from the twentieth into the twenty-first century, connecting Vietnam, the Gulf War and the Balkan Wars with the more recent and extended wars in Iraq and in Afghanistan, dominating how wars and their conduct are understood. This has shaped Western perceptions of and sensibilities to modern war in terms of its causes, consequences and victims as remote, as other and as distant spectacle.

The importance of television to war became apparent during the Vietnam War. In particular, Vietnam established an enduring political belief that television news coverage undermined public support for the US military campaign and was one of the reasons why America lost the war. Subsequent research in media studies shows us that the media was not as influential in framing domestic opinion about the war. Instead, only a small percentage of film reports on television news during the conflict depicted actual fighting and graphic scenes of the dead or wounded (Braestrup 1983; Hallin 1986). From the military's point of view, however, failure in Vietnam led America's armed forces to avoid thinking about counterinsurgency in the thirty years preceding the 2003 invasion of Iraq (Nagl 2005).

The belief that television coverage influenced the public to turn against US engagement in Vietnam shaped how American commentators and politicians framed and legitimised wars in the Gulf in 1991 and Iraq in 2003 (Hoskins 2004). Thus President George H. W. Bush in his televised speech to the nation announcing the beginning of the air attack on military targets against Iraq at the start of Operation Desert Storm declared 'that this will not be another Vietnam'. Bush's subsequent lines echoed his earlier comparison to Vietnam, stating that in the execution of the 1991 Gulf War '[o]ur troops will have the best possible support in the entire world, and they will not be asked to fight with one hand tied behind their back. I'm hopeful that this fighting will not go on for long and that casualties will be held to an absolute minimum' (cited in Hoskins 2004, pp. 34–5). For the Bush administration, the depiction of blood on television restricted American forces from doing their job, and so the technologies, tactics and media operations of the United States needed to be adjusted to avoid alienating the US public from the realities of warfare.

By the end of the campaign, however, it was clear that Bush and his administration need not have worried about drawing comparisons between the 1991 Gulf War and Vietnam. Most US news networks – and many broadcasters around the world-dependent upon the pool reporting system for content – fully embraced the administration's efforts to sanitise the visage of war. This made it easier to set out narratives that reinforced the studium of war, where politics, the armed forces and the media's representation of the conflict were in harmony. Thus, Desert Storm typified politics by other means. The media limited the violence of war to its political framing. This was repeated by the armed forces, where daily briefings focused on smart, 'laser-guided' weaponry combined with endlessly looping video taken from missile nose-cone cameras and satellite imagery.

This helped the military's media handlers portray the war as a precise, seamless and bloodless conflict, fought cleanly and at distance. As the bloodless depiction of America's air assault on Iraqi efforts to flee Kuwait City on 26 February 1991 highlights, the goal was to avoid alienating American viewers watching news coverage on the 24/7 news networks by showing them burnt-out vehicles rather than burnt dead bodies.[8] However, even this coverage demonstrated the limits of media management and instead led military planners to think carefully about how to apply force as the coalition headed towards Iraq proper.

The Gulf War in 1991 sustained the belief in the centrality of television news as a shaper of political decision-making. However, the war also drove a new wave of writing on how 24/7 real-time news reporting shaped political and military decisions in an effort to understand the 'CNN effect'. For journalists like Nik Gowing, real-time reporting affected political rhetoric but rarely changed the policy of presidents and governments (Hoskins and O'Loughlin 2010). For others, the focus shifted towards the mediatisation of war and the role of television in reframing perception. That very little happened most of the time was not an issue. There was a sense in which the whole world was watching. This created a shared experience, irrespective of whether you were watching from the Whitehouse or from the streets of Baghdad. The result left viewers experiencing the immediacy of war and the expectation that something was about to happen. This made for compelling viewing and certainly made CNN's reputation. And it was this, as a kind of new perception of war, that really established the scholarly field of war and media, where the un/reality of the televisual spectacle could be theorised by philosophers like Jean Baudrillard (1991/5), Paul Virilio (1991) and McKenzie Wark (1994).

If the Gulf War reinforced established narratives about the political purpose of war, then the genocide in Rwanda and Burundi and the massacres that followed the collapse of Yugoslavia represent narratives that Western media struggled to reframe. During the Bosnian War, for example, the world watched in anticipation of some sort of intervention by European or American armed forces even as the international community sat back in a state of paralysis. Not even daily news updates from the four-year siege of Sarajevo could bring about political action from the West. This should have ended any kind of presumption that television news might bring about political action. Instead, as David Rieff writes in his book *Slaughterhouse*,

> 200,000 Bosnian Muslims died, in full view of the world's television cameras, and more than two million other people were forcibly displaced. A state formally recognised by the European Community and the United States ... and the United Nations ... was allowed to be destroyed. While it was being destroyed, UN military forces and officials looked on, offering 'humanitarian' assistance and protesting ... that there was no will in the international community to do anything more. (Rieff 1996, p. 23, cited in Keenan 2002, p. 104)

A greater concern was that the daily diet of violence being broadcast on a 24/7 basis left some scholars worried that television didn't so much discourage societies from fighting but rather led them to enjoy the spectacle of it. Thus, while Tony Blair could plead that military intervention in Kosovo in 1999 constituted a just war, his Chicago Speech also established the basis for engaging in humanitarian wars all over the world.[9] In this new context, then, professor of comparative literature Thomas Keenan raised the question of '[w]hat if, because the cameramen and the images were there, and because they are supposed to make a difference simply by virtue of what they showed, the disaster continued?' (Keenan 2002, p. 340). If this was the case, then genocide and massacre represented the punctum of war, where violence occurred for the sake of violence and war had lost touch with its political reasoning.

The Studium and Punctum of War Post-9/11

Up until 09:05 Eastern Time on Tuesday, 11 September 2001, the studium of war dominated the way war was understood. However, as George Bush was left frozen in front of the press pool and second graders at the Emma E. Booker Elementary School in Sarasota, Florida, the punctum of war properly shredded the conventional understanding of war. In those few moments before Chief of Staff

Andy Card whispered into the president's ear that 'America was under attack', the entire civil–military bureaucracy of the United States was revealed to be inadequate to the threats Americans now faced.[10] As the decade progressed, it became clear that this was not a blip in the studium of war. Rather, this moment of disruption would become the norm in the representations of war in the twenty-first century.

Thus, the narrative ebb and flow around the studium and punctum of war came into more obvious relief during, and in the immediate aftermath of, the live televised terrorist attacks on the Twin Towers and the Pentagon on 9/11 in 2001. Representing a peak in TV spectacle, these attacks reinforced television's persistence as the media of choice in early twenty-first-century depictions of war. At the same time, the spectacle and horror of the attacks brought the chaos of war directly into American homes in a way that genocides in Rwanda, Burundi or Yugoslavia never could. Consequently, in the days and weeks that followed, the media struggled to compose a framing of 9/11 that anchored interpretations of what Americans had experienced when they turned on their televisions that morning. The immediate framing device that the media and politicians chose drew on references to Pearl Harbor in 1941. However, as the subsequent invasions of Iraq and Afghanistan became more intractable, the possibility that the GWOT would neatly fit into the studium of war soon evaporated.

That conventional narratives about war had somehow reached a breaking point is best illustrated by President George W. Bush's televised 'Mission Accomplished' address on 1 May 2003. The speech was delivered after the fall of Baghdad, with the military establishment telling him that the war was over. Bush's speech is remembered for two reasons. One is his assertion that this battle was 'fought for the cause of liberty, and for the peace of the world'.[11] The other was that within the year it was crystal clear that the Americans could not keep control of a growing insurgency in Iraq.

The initial decisions to invade Iraq were justified in terms underpinned by Clausewitzian thinking. In 2002, however, the Bush administration had already decided to ignore the 1949 Geneva Conventions to respect the rights of al-Qaeda as legal combatants in war. Instead, al-Qaeda would be treated as unlawful combatants and if captured would not be treated as Prisoners of War. The Bush administration established a new legal basis for the GWOT, justified on the grounds that this terrorist organisation did not respect the laws of war or respect distinctions between combatants and non-combatants. This ignored conventional twentieth-century wisdom about war and heralded the descent into the chaos and disorder during the occupation of Iraq. This was most obviously emblemised by the human rights abuses by American soldiers of Iraqi detainees at the Abu Ghraib prison in 2004.

As countless documentaries of US occupation attest, under these new conditions, sustaining conventional establishment narratives in the media was not possible.[12] The chaos itself was a function of American failures to properly prepare for occupation and security in Iraq and an assumption that the Iraqis would be pleased that democracy would now be possible. As *New York Times* journalist Dexter Filkins observed, the problem was that '[t]here were no Army manuals on how to set up a local government ruined by 30 years of terror, no maps for reading expressions on the face of a Sunni sheik, no advice on handling the city engineer who was taking bribes to dole out electricity'.[13]

The inability to manage the technical aspects of running a society was only matched by the lack of understanding that Americans had of the complexities of Iraqi politics. Consequently, the Americans could start the campaign with good intentions, trying to build bridges between themselves as occupiers and the Sunni and Shia communities they were left to police. However, as looting, criminality and score-settling turned into more serious insecurity, the media found themselves having to tell two stories about Iraq. One side of the news story came from the Coalition Provisional Authority (CPA) press briefings where Paul Bremmer, the head of the CPA, would explain that essential infrastructure was being rebuilt and services stood up. The other side of the news story were the obvious signs of chaos and disorder all over the country. While the Whitehouse claimed that they could not foresee the disorder that emerged following occupation, the *New York Times* described the administration's preparations as a 'Blueprint for a Mess'.[14]

Emblematic of America's strategic failure was the CPA's decision to remove Saddam Hussein's political acolytes from positions in the Iraqi government and Army. This de-Baathification decision alienated the Sunni community, who were the predominant beneficiaries of Saddam's use of the Baath Party for managing religious divisions in Iraq. By disbanding the Iraqi Army, the Americans 'made 450,000 enemies on the ground in Iraq'.[15] These ex-servicemen had some weapons proficiency and without jobs could not support their families. As sectarian violence erupted across Iraq over subsequent years and as Americans responded to that violence with yet more heavy-handed tactics, the press struggled to tell a story that would fit into conventional narratives about war. The result was a spiralling descent into chaos as those trying to impose order found themselves repeatedly outmanoeuvred by those willing to apply even more deviant forms of violence. The American failure to attend to social and religious divisions within Iraq seeded the civil war that went on to dominate Iraq for the next decade and a half.

When it comes to discussions about war and media, however, the political and military decisions represent only one aspect of the story. Of equal relevance are the changes that were going on in relation to media technologies and broader information infrastructures. In this respect, it is important to note how the broadcasting era associated with MSM was increasingly struggling to dictate how news would be reported. Before the era of the smartphone and social media, the key medium for conveying messages that did not conform to the studium of war was the weblog. Here Michael Yon, a former Green Beret with US Special Forces, played an important role.[16] Offering counter-narratives to official reports or from journalists embedded with the military, Yon's blog was particularly important for publishing the punctum of war, revealing an impression of war that MSM was otherwise unwilling to discuss.

By 2010, blogs were beginning to be displaced by the smartphone and social media. Suddenly users could broadcast content over the web without needing to go through a traditional MSM broadcaster. This made it possible for anyone to participate in publishing counter-narratives to those offered by governments or in the broadcast media. In effect, the punctum of war had become the norm as opposed to the transgressive, framing how subsequent wars would come to be understood. But while the glare of social media renders an experience of war as always immediate, arresting and spectacular, driven by the algorithms of attention-dominance, this distracts from or covers over other forms and places of war. Radical War, as we introduced it above, is 'radical' not only in the scale and effects of the interconnected transformations in media, technologies and infrastructures but also because these appear to bring war into the open while at the same time providing cover for its operations.

If the punctum of war is now the norm rather than the exception, then the war in Ukraine constitutes a good example of where Radical War now exists. For it is in Ukraine that the hype of connectivity and contagion, that long-held assumption in media studies on the relationship between knowledge and action, fails. The logic of the 'connective turn' (Hoskins 2011a, b) – the sudden abundance, pervasiveness and immediacy of digital media, communication networks and archives – is most revealing in its fissures, its gaps and its silences.

The Russian war on Ukraine was undeclared, the enemy's identity opaque and the conflict fought by soldiers without uniform. But this is not war under conditions of broadcast-era media. Rather, this is an overt/covert war: seen but not seen, known but not known. Rather than illuminating this war and its causes and consequences for a watching world, the continuous penetration of dense digital networks, social media, television stations and press, whether global, regional or local, has been a source of obfuscation. This opacity has in turn become war's most vital weapon.

For example, an integral part of Russia's military campaign against Ukraine and Crimea has involved the use of a troll army to spread fear, propaganda and disinformation (Pomerantsev 2019). In part, this has involved setting up fake social media accounts and luring Ukrainians into disseminating fake news to create confusion.[17] Thus, in an interview, a former press officer for a Ukrainian paramilitary group told the story of how their cameraman, Artem (who was vulnerable and suffering from PTSD following his experience of filming in battle), had been manipulated by Russians employing social media. In this example, a Russian Federal Security Service

(FSB) employee set themselves up as a moderator of a Facebook group, creating fifteen new accounts (a mix of human and social bots), which looked entirely benign and included the ubiquitous social media meme that is #emergencykittens. These accounts added Artem to their new group and then liked some of Artem's posts and wrote friendly and supportive messages.

Over time, the group gained Artem's trust to the extent that he then reposted material that was even critical of the paramilitary cause he supposedly worked for and ultimately led him to be fired from his job. The principal problem with Facebook, according to the press officer recounting this story, was that the material was 'shared too quickly, without any analysis'. The FSB moderator's trustworthiness was effectively enhanced by a critical mass of other members or connections and the follow-up process of liking, linking and sharing. Recruited in slow time and by what appeared to be an authentic and legitimate group, Artem was undone by sharing links at speed and on the basis that those supplying him information could be trusted.[18]

Through this process, the machinic and the human work in concert to create what is in effect an alternative reality, an in/visible war. Thus, as Nadiya Romanenko et al. explain:

> First, a troll posts something. Then a host of other trolls begins to like it, comment and repost. The Facebook's algorithm perceives it as interest of living people towards this post, which begins to appear in news feeds of NON-trolls who happened to befriend a troll. If any post is successful, regular people start to repost it – and then journalists as well. The same scheme boosts video clips on YouTube.[19]

And here we have the triumph of digital attention and 'like' war (Singer 2018). Information is laundered through multiple bot and real accounts.[20] This can be used to amplify propaganda in ways that afford the troll armies the cover of legitimacy even as they provoke social media users to engage with what is trending (Patrikarakos 2017).

But this is not only an informational war between competing groups or enemies. On the contrary, Ukraine has been seriously damaged by a toxic media ecology in which the media is at war with itself. In the words of Natalia Ligachova, editor-in-chief of the NGO Detector Media: 'There is a war between those journalists who support the government and those who criticise it. As a result, people no longer believe anything … There is informational war within the media community especially in social networks, and it weakens us.'[21] The great promise of the digital revolution in engendering the values of open access and garnering the power of mass scrutiny by the multitude was a fallacy. Instead, the connective turn has delivered an environment in which society can no longer glimpse a coherent image of itself in order to arrest its conflicts, or as Yuri Makarov, a Ukrainian TV presenter, said in an interview in 2017, '[the media] … look like a reflection in a broken mirror'.[22] As a consequence, interpretations of the war in Ukraine are stuck in what appears to be a permanent loop of counter- and transgressive narratives that leave the studium of war constantly trying to catch up with war's punctum.

The New War Ecology

War and the capacity of individuals to participate in it have expanded dramatically. Now everyone can be photographer or a journalist, recording incidents and filming events that can be published online and broadcast to the world. More than this, however, the simple act of carrying a connected device that can share geo-coordinates means that anyone can become a sensor and unwittingly play a role in war. Of course, people engage with war through social media, watching and amplifying events and recontextualising them against the historical record and in relation to people's memory. However, as we have argued, this has reversed the broadcasting model that emerged during the twentieth century, creating a web of producer/consumers who are relentlessly sharing, promoting and contesting different narratives about war. This has created as many individual narratives about war as there are users of digital devices engaged in discussing it. In light of the quantity of material that is now being produced, neither social media companies nor the state can easily control what is being published and said online. Even in China, a country that has done more than most to suppress the punctum of war, it has not been possible to prevent transgressive perspectives from shaping discourse on political violence.

Given the scale and complexity of the changes that have taken place, we argue that it is not possible to locate these emergent forms of warfare within existing models of representation and ways of seeing the world. Put simply, the quantity of data that now exists disrupts epistemic frameworks that rely on stable narratives born out of what Émile Durkheim would have described as a social fact. Instead, technologists and states are forced into using sophisticated AI algorithms to try to make sense of the growth of data. These tools are not readily understandable and so leave ordinary users in the dark about the patterns of data that are influencing what they see online. Consequently, user sense-making is partial and is easily caught in the information prism of online life. As whistle-blower Frances Haugen explained, this is easily exploited for profit by social media companies like Facebook.[23]

War today is radical in that it has moved into a fourth dimension, contorting the studium and punctum of war in which the old order is still oddly present and continuing but at the same time doesn't appear to make sense under the conditions of the new. And the very move into this dimension in itself obfuscates Radical War, not least because conventional narratives about war can still be found, jostling for attention with those in the fourth dimension.

War in this dimension nevertheless represents a fundamental break with war before 9/11. This is in part a function of the scale of data and its ability to be accessed on a nearly instantaneous basis. More importantly, however, the availability and use of smart devices ensure that billions of individuals participate in war whether wittingly or not. Digital devices can be geolocated. Their microphones and cameras can be switched on remotely. This creates an accessible archive that can be mined for information so that social connections might be identified. Even if someone is not directly engaged in filming, recording or commenting on war, the potential for their device to act as a sensor remains. Participative war thus heralds a break with models of political violence that implied there was a difference between victim, perpetrator and bystander (Hilberg 1993).[24]

Diagram 2: The Hierarchy of Violence

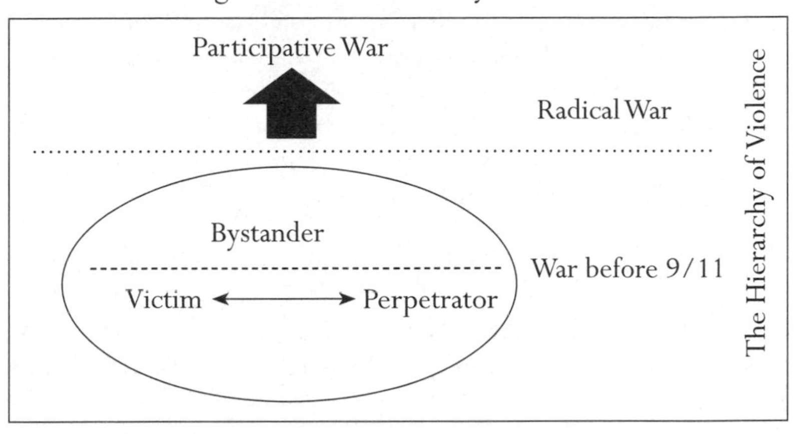

In contrast to this approach to political violence, Radical War establishes a hierarchy of violence where smart devices effectively guarantee participation in war. As a result, the relationship between victim, perpetrator and bystander all collapse into the category of participant. The removal of this middle ground has important but poorly understood implications for how combatants and civilians are understood on the battlefield, in the civil–military bureaucracy and in relation to how military power in the twenty-first century is generated.

Even more corrosively for those who hold on to the Clausewitzian theory of war, the hierarchy of violence collapses civil–military distinctions as they are framed by *On War*. One of Clausewitz's central challenges in *On War* was to account for the passions of the people as they related to the rationality afforded by government and the calculated preparations of the armed forces. For strategists in the early 1800s, this was an urgent and important challenge, because as the historian Peter Paret observes, war had been taken 'out of the hands of a relatively restricted elite commanding long-serving professionals and made [into] the business of the people'. Without a framework for understanding how to harness the passions of the people, the conservative monarchies of Europe would have been unable to defeat French Revolutionary and subsequently Napoleonic armies (Paret 2004). Clausewitz considered the problem in Book 6, Chapter 26 when he said:

> A people's war in civilised Europe is a phenomenon of the nineteenth century. It has its advocates and its opponents: the latter either considering it in a political sense as a revolutionary means, a state of anarchy declared lawful, which is as dangerous as a foreign enemy to social order at home; or on military grounds, conceiving that the result is not commensurate with the expenditure of the nation's strength.

Having identified the problem, Clausewitz squared the dilemma of how to account for the place of the people in war by constructing a trinitarian model where the people, government and army were separate but related in the efforts to direct war towards its political purpose.

In the twenty-first century, the hierarchy of war collapses Clausewitz's trinitarian model. The passions of the people are diverted and warped in the digital prisms of the fourth dimension. Everyone participates if they are connected to the web. This removes the civil–military divide that frames the way war before 9/11 was understood. The evolving punctum of war complicates the process by which social consensus about war can be reached. This does not prevent governments and armed forces from defining the political objectives of war but does open up questions about how broader society is harnessed by the state. Radical War implies the demise of the civilian, such that everyone becomes a vehicle for realising political objectives irrespective of their combatant status. IS does not recognise the distinction between combatants and civilians (Ingram, Whiteside and Winter 2020); everyone must choose a side.

Radical War thus comes with a number of axiomatic features that are a function of contemporary information infrastructures. This starts with the notion that everyone can record and publish their experiences of war. Although broadcast media continue to operate in the twenty-first century, their relative power is disrupted by users producing their own recordings and narratives. The splintering of opinion to the level of the individual user of media – the participant – complicates the process of formulating and generating consensus about war. The nature of connected devices further complicates these issues by making it possible to harvest and weaponise the data and metadata that they produce. This creates a hierarchy of violence that collapses civil–military divides in ways that have yet to be fully mapped and understood. It is also clear, however, that beyond these essential characteristics, Radical War also implies a number of features that shape the relationship between data, attention and control. As these variables come up across the chapters that follow, we list them below.

1. The Digital Individual

The hierarchy of violence enables everyone to participate in war. It also destroys existing categories of identity. Victim, perpetrator and bystander disappear as credible markers of someone's role in war. Nor is it possible to hold on to the categories of soldier and civilian. Instead, identity has collapsed into the category participant. These participants may or may not be intentionally engaged online. However, anyone who has registered with an internet service provider (ISP), who owns a smartphone or who has a social media profile is a digital individual. Whether the digital individual and the identity of the person are the same is not guaranteed. And today we can see identity and anonymity as part of a digital spectrum along which individuals can hide. Denial, fakery and so-called 'post-truth' thrive here to constantly cast uncertainty, which is at the heart of social media prisms.

2. Flattening Experiences of War

The relationship between war and political violence mirrors the collapse of the categories of soldier and civilian. This flattening of experience effectively means it is possible to talk of war as political violence and political violence as war. For scholars wedded to Clausewitz, this collapse of distinctions will be unacceptable. Whether it is possible to forget Clausewitz given his significance for those engaged in the study of war is an open question (Öberg 2014). However, if we allow for the possibility that practice will dictate the extent to which a theory has explanatory power, then we have at least to accept the possibility that the world is not as Clausewitz says it is. In this respect, the digital individual's experience of war in the twenty-first century cannot easily be accounted for by a trinitarian model of war that came out of the nineteenth century. This implies that we must look beyond Clausewitz for our explanation of war and political violence.

3. Everywhere War

Following the coordinated terrorist attacks in Paris on 13 November 2015 that killed 130 and wounded hundreds more, President François Hollande declared that 'France is at war': 'We are in a war against jihadist terrorism, which threatens the entire world.'[25] Picking up the baton the following week, the French philosopher Bernard-Henri Lévy described this as a new kind of war: 'So it's war. A new kind of war. A war with and without borders, with and without states, a war doubly new because it blends the nonterritorial model of al-Qaeda with the old territorial paradigm to which Daesh has returned. But a war all the same.'[26]

What we do not want to suggest is that war will be any more or less kinetic than it already is. However, the spaces that war can now reach stretch well beyond the battlefield itself, even as war can still have a physical, geographical location. This is simply a function of the widespread availability of information infrastructures transmitting data about war, and distracting from other forms of political violence, in ways that have never previously been possible.

4. Weaponising Digital Divides

The new war ecology is not evenly spread across the globe. Different countries, organisations and armed forces have radically different information infrastructures. Consequently, the data trajectories of the new war ecology vary. These infrastructures determine whether data will move more or less efficiently (Lewis 2014). Where access to fast speed data transfer is limited, bottlenecks may occur. Alternatively, bottlenecks may be created by actors interfering in existing information infrastructures in an effort to limit data transfers. The value of these bottlenecks lies in the way they can be exploited for informational advantage, shaping and redirecting attention in ways that can be useful to those trying to divert narratives. The attention afforded to some killings and not others hints at a wider shift in terms of the anticipated relationship between the scale of war or human suffering, the extent of its reporting, and the structural features of contemporary information infrastructures. This is particularly noticeable in the uneven evolution of the new war ecology where local and global media ecologies collide and/or intersect.

At their most elementary, digital divides might be exaggerated by offering different data transfer rates to different communities. At a more sophisticated level, digital divides are created and sustained through the creation of a sovereign internet. Equally, the capacity to lay down a temporary network infrastructure that provides access to cloud computing also ensures that data flows can manipulated, and information asymmetries exaggerated in an effort to control narratives. Similarly, bundling network services in with the purchase of a smartphone may leave users tied to networks without access to a varied media ecology. Finally, in highly sophisticated media ecologies, digital divides can be manipulated with spoof voices and deepfake video files.

5. Data-saturated War

The challenges produced by the current evolution of the information age stem in large measure from the huge quantities of data that have followed a process of datafication. In media studies, this is seen as part of a process of what Nick Couldry and Andreas Hepp call 'deep mediatisation', namely a 'wave of digitalisation and datafication' that equates to a 'much more intense embedding of media in social processes than ever before' (Couldry and Hepp 2017, p. 34; Hepp 2019). But what is distinctive about today's media age is that this embedding is no longer a matter of our reliance upon devices and networks, but rather a dependence. These circumstances create the conditions for a connective turn such that the entire epistemological framework for understanding the world is deeply mediated. Even as it does this, the datafication of the everyday produces a mediatisation of war that collapses the categories of war and media into each other, feeding processes of amplification in social media and leading to the 'like' war phenomenon (Singer 2018). At the same time, the availability and abundance of information about war ensures that connected societies can engage with the atrocities of war with an immediacy that has not been experienced at any previous point in time. The result has not led members of the UN Security Council or the societies they represent to demand more military intervention to protect human rights. Instead, it appears to feed an appetite for spectacle.

The flattening of media and war into one register unsettles the historian's traditional role in shaping a social version of events, which in turn has implications for how we engage with memory and the commemoration of war. The information philosopher Luciano Floridi conceives of this as 'a new threshold between history and a new age called hyperhistory' (Floridi 2013, pp. 37–8). He argues that

> human evolution may be visualised as a three-stage rocket: in prehistory, there is no Information Communication Technology (ICT); in history, there are ICTs, they *record* and *transmit* data, but human societies depend mainly on other kinds of technologies concerning primary resources and energy; in hyperhistory, there are ICTs, they record, transmit and, above all, *process* data, increasingly autonomously, and human societies become vitally dependent on them and on information as a fundamental resource. (Floridi 2013, p. 38)

It is in these new circumstances that narratives about war accelerate and clash, leading to new instabilities in how war is understood.

6. Accelerating Memorial Discourses

The new war ecology is not just about the present but also crucially recognises how media, memory and history exist in a new nexus of unprecedented complexity and scale, framing participation and capturing attention. How war is perceived, experienced, won and not won, legitimised, declared, fought and lost, studied or ignored, hidden and made visible for different actors, for different ends, in and over time, is entangled with remembering and forgetting. Datafication makes ambiguous the relatively steady unfolding of these relationships and the certainties and stabilities that traditional processes of historical research once offered. This transforms narratives about war, reframing how it is understood by its actors, unsettling accepted history and supplanting it with a memorial discourse that stands independently of historical truths.

But Radical War is paradoxical. It is fought and experienced through a radicalisation of memory, with the contested past shaping what war is in the present. And the digital present has come to take on ever greater memorial force, offering a refuge from the insecurity and incomprehensibility posed by today's wars. In this context, the velocity and volume of incoming media feeds can only be made sense of within a schematisation of what warfare looks like. Making sense of war in memory thus evolves at a pace that cannot be fixed but is a dynamic, imaginative exercise that is directed and shaped from the present and must also fit within schematic frameworks the mind forms from past experiences and commonly received understandings.

Thus, what is and what will become of the memory of recent and unfolding war is much more immediately contested by an array of actors through the abundance and accessibility of data and information about those wars in the participative digital sphere. And these trends unsettle how memory is reconstructed and how historians can defend their role in an environment in which 'the past becomes what Everyman chooses to accept as true' (Lowenthal 2012, p. 3). In the prism of social media, attaining a settled version of events given the relentless posting and mixing of opinion, information and disinformation, stretches historical method (Hauter 2021).

7. The Hacked Image

The digital image cannot be kept secure. Now that soldiers are on the digital global grid, digital imagery is found in campaigns that require some kind of visibility or disguise. As with the 'little green men' who fought Russia's proxy war against Ukraine, these are subject to the risks of inadvertent as well as intentional exposure. And this affects everyone who makes use of the web irrespective of where they are from. Thus, American soldiers in their ambition to keep fit, upload figures from their Fitbit devices to the cloud and in the process expose their geolocation data to adversaries in what amounts to significant breaches of operation security.[27] Equally, selfies taken by Russian soldiers helped Ukrainians and international investigative communities like Bellingcat destroy Russian propaganda efforts that claim 'there are no Russian soldiers in Ukraine'. Similarly, Bellingcat debunked claims that Russian Military Intelligence were not behind the attempted assassinations of Sergei and Yulia Skripal in the city of Salisbury.[28] By 2019, these failures led the Russian State Duma to pass an 'Anti-Selfie' law forbidding military personnel from carrying or using internet-connected devices to share information about their service.[29] No doubt elementary leaks will get plugged, but given the ongoing march of the IOT and the networked battlefield it is clear that the deeply mediated digital world will continue to provide new vectors for hacking data and manipulating it for propaganda purposes.[30] Radical War, then, requires a lens that highlights the complex intra-dependent relationship between war propaganda, smart devices and web platforms while also accounting for the transformed role of the image in sustaining and undermining military campaigns.

Making Sense Out of Radical War

How, then, can we begin to make sense of what we are calling a new war ecology? We use this term to argue for a holistic way of imagining war, media and society such that we do not lose sight of or isolate any of its interconnected elements, experiences and effects. Instead, we argue that Radical War is infused by data and is fought and experienced, revealed and obscured within and by a new knowledge base or 'information infrastructure' (Bowker and Star 2000). To reveal this ecology is to rethink the parameters of war, to incorporate its human and non-human elements, to expose the appropriation of its information infrastructures, and to interrogate media non-effects as much as claimed effects.

So, as we engage with the idea of Radical War, we need to reflect on how smart devices and contemporary information infrastructures reframe participation and remake the relationship between war, society and the media. In the chapters that follow, we start to map out where the principal challenges in this new nexus might lie in relation to the organising principles of data, attention and control. We recognise that the agenda we set out is much wider than we can document here. Despite this, the essential characteristic of Radical War is framed by the notion of a hierarchy of violence between combatant and non-combatant. This hierarchy ensures that data and metadata is available for accelerating war but also for framing how new targets are identified and attacked.

What is also apparent is that datafication will significantly complicate the process of establishing consensus about the place of war in society. This unsettles the historian's traditional role in shaping a socially acceptable version of events and makes memory even more significant in framing attention. The abundance of data also leads technologists themselves into positions of power, enabling them to take even greater control over aspects of society as their systems displace the capacity of governments to administer their sovereign geographies. This has important and yet under-explored ramifications for civil–military relations that we can only point towards here. Finally, the new war ecology is itself in a permanent state of flux as new technologies and experiences shape and reshape the narratives of war. This needs attention if we are to understand war in the twenty-first century.

PART 1

DATA

Diagram 3: Data

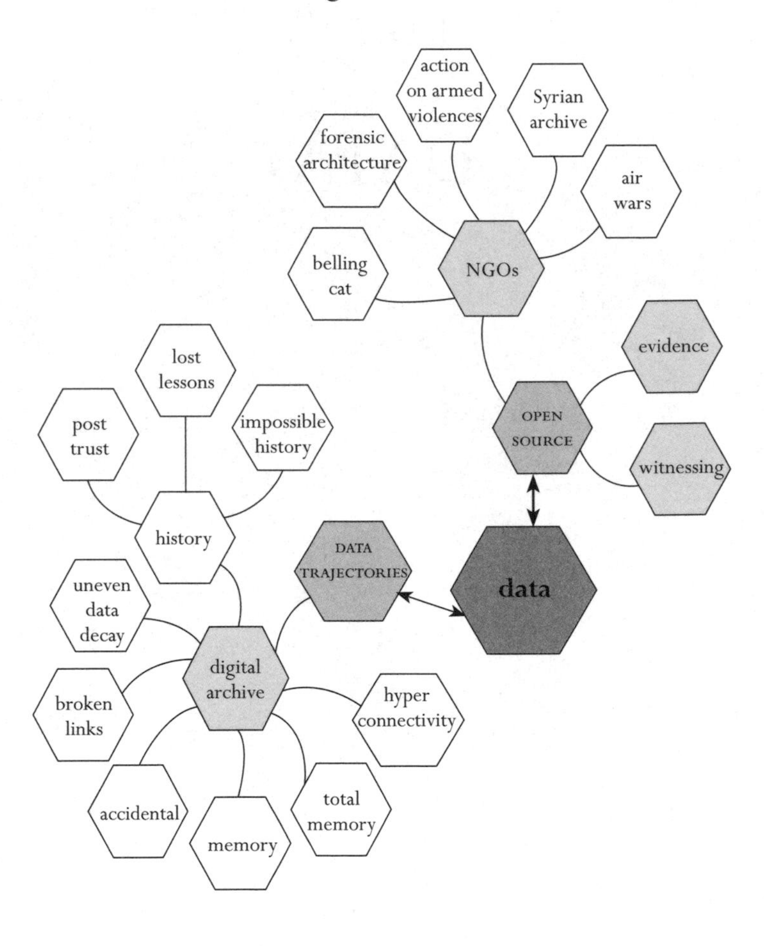

3

THE RUPTURED BATTLEFIELD

On Thursday, 13 May 2021, MSM reported that the Israel Defence Forces (IDF) were massing on the border of the Gaza Strip. The IDF were apparently preparing to invade and then occupy Gaza.[1] The next day, journalists received a WhatsApp message claiming an invasion had begun. As the day unfolded, however, it became clear that troops had not entered Gaza and instead an air campaign had been launched. According to some commentators, this was part of an IDF disinformation campaign. The apparent goal was to persuade Hamas to concentrate its fighters in a series of underground tunnels in preparation for combat. This would then present an ideal target for the IDF air force. As Nir Dvori, a military reporter for an Israeli news network subsequently explained: 'Then for 35 minutes, 160 planes hover over Gaza and drop 450 bombs, which are over 80 tons of explosives, on the entire Gaza "metro".'[2] Annoyed journalists subsequently complained that the media had been deployed as part of the IDF's military campaign. The MSM had apparently been duped into helping the IDF destroy Hamas's fighters.

The idea that armed forces might use MSM to shape the battlefield is not new. In the new war ecology, however, managing the media is not a simple case of talking to journalists. Indeed, even as MSM were being played, the IDF also asked Israelis living near the Gaza border to disconnect their webcams.[3] The goal was to deny Hamas the opportunity to hack these devices and use the feeds to help identify targets or generate propaganda. With this request, the IDF implicitly acknowledged that they could not prevent these systems from being taken over by Hamas and used against Israel. Instead, they decided to intervene more directly, asking Israelis to switch these devices off.

As the campaign against Hamas unfolded, the IDF undertook air strikes against several high-rise buildings they claimed contained Hamas fighters. One of the buildings they destroyed was also the home to mainstream news agencies the Associated Press and Al Jazeera.[4] The IDF sought to justify attacking the offices of two international news media companies on the grounds that their buildings were also home to enemy combatants. Attacking these buildings also had the net effect of hampering reporting in a way that the IDF hoped would further shape the media environment.

On the one hand, by attacking the offices of two major news networks the IDF sought to control the speed at which broadcast media might convey stories from the Gaza Strip. On the other, the IDF could not control information revealed by hacked devices. As *New Yorker* journalist and Tel Aviv resident Ruth Margalit observed, the IDF were no longer able to control how violence might be presented within and beyond Israel:

> This was the first time you saw social media rearing its head … you saw the fraying apart of the omnipresent military censor in Israel. Up until that point all the news within Israel but also the foreign news outlets had to pass through the military censor. And suddenly because of social media and because of these other internet journalism [*sic*] … the military censor has lost its place. And I think in Israel, the Israeli Army and the Israeli Government are certainly grappling with that. How do you control the narrative and keep certain stories from being aired? And the truth is now that they really can't.[5]

Margalit's statement points to the problem that armed forces now face. People participate in war by virtue of their connected devices. Military activity and its representation in the twenty-first-century media-verse are deeply intertwined, which is shaping how armed forces are preparing for war. Now armed forces must have the capacity to fight at the speed of data transfer. This will allow them to be ahead of the media narrative, shaping stories even before journalists have had chance to broadcast. However, the implications of this have yet to be properly thought through. In this chapter, we examine where these ideas have emerged from and explain why fighting amid the velocity of informational trajectories about war is a problem, especially in relation to making sense of and learning lessons from war.

***

In the twentieth century, battlefields were located geographically somewhere in the world. By the mid-2000s, with the emergence of 24/7 broadcast media, there was according to British Army General Sir Rupert Smith a 'theatre of war' (Smith 2006) that continued to occupy a defined space. Mirroring the IDF's efforts in relation to Hamas in the spring of 2021, the test was how to manage the story the armed forces repeated to their new TV audiences. Over the second decade of the twenty-first century, the challenge faced by the military has changed. Now virtually everyone can broadcast what is happening via their smartphone. This new data-saturated battlefield spreads information across an arc of trajectories stretching from the immediate environs of the battle itself, all over the internet and across social media platforms. As a result, the battlefield has extended beyond a physical space and can be found located in multiple contexts all over the world. What's more, this data can be read, rewound, reused and recontextualised over and over and in ways that cannot be predicted. The multitude of narratives that this creates and the asynchronous way in which they appear has upended Smith's theatre of war even as it obscures war's political rationale.

For the military, this accelerated information environment makes it hard to learn lessons. If military operations have to keep up with the speed of information as it is spread across civilian data networks, then when does war stop? If war is always read, rewound, reused and recontextualised, what chance is there to reflect as part of a post-operation report? Processes of digitalisation are challenging the state's capacity to manage its archive. Processes of datafication imply the truths of war are only going to become more muddied in the digital prisms of the new war ecology. This has serious repercussions for how military organisations and society generate meaning from history.

Now that war is participative, civilians feature on the battlefield not just as keyboard warriors, human shields or collateral damage but also as unwitting participants in war. In the case of the IDF, connected devices provide global positioning coordinates and whatever material people are recording or tagging online. This extends the military's battlefield, providing data that can in turn be weaponised and used in the construction of target lists. Depending on the observer and the civilian's location, someone's legal rights may have been circumvented (McDonald 2017). In this case, they cannot be classified as bystanders but instead they have become victims or perpetrators.

Not only has this created uncertainty over the effects that future military technology is capable of delivering in battle, it has even upended the military's capacity to frame the political messages that might be inferred from the use of force (Rid 2007; Briant 2015a, 2015b). This presents a very real problem for armed forces, for we are no longer in the realm of discussing military effectiveness without, in the same breath, also talking about audience effects.

This challenge is only made more complex by the speed of the data trajectories themselves. For data is generated by a variety of actors, both inside and outside the armed forces. These military and civilian actors have differing needs, make use of different information infrastructures and engage their audiences differently. Compartmentalisation and security classification guarantee that military information infrastructures are slower than those civilians working on open networks. Obviously, this affects who sees what and when. This also tells us something about the unevenness of the new war ecology. At the same time, we start to get a glimpse of the 'deep mediatization' (Hepp 2019) of everyday life and the corresponding interpretive frameworks that shape the military's understanding of war.

The result collapses distances and boundaries, producing asynchronous experiences of war and recreates war in online spaces that compel adversaries to battle each other in the fourth dimension. By de-territorialising war, we find that it emerges everywhere, folding memes, images and narratives so as to transgress the national boundaries of the analogue world and the physical site of any particular conflict (Gregory 2011; Singer 2018). Victory in this context is not about decisive action leading to conclusive results but to some appears to be analogous to forms of limited war (Stoker 2019) and to others as a process of managing violence (O'Driscoll 2019). At the same time, digitalisation is feeding and expanding a cycle of thinking in which the classical categories of war and peace, combatant and civilian, inside and outside the state (Walker 1992) no longer serve to help us understand the breadth of changes Radical War has prompted.

These issues are set to multiply as technologists introduce an MIOT (see Appendix) directly on to the battlefield itself, complicating existing patterns of datafication, systems integration and battlespace management even as it seeks to accelerate warfighting. As this focuses attention towards intelligence collection, analysis and targeting, fighting is only as relevant as the capacity to process data for the purposes of identifying targets (Nordin and Öberg 2015). The increased focus on data reveals a number of organisational and bureaucratic disjunctions. What becomes clear is that data trajectories are faster on the battlefield but slower within wider and typically older government infrastructures where more considered approaches to decision-making are required. At the same time, participants out of uniform and unhindered by security protocols will continue to make use of civilian information infrastructures to disseminate data from the battlefield even more quickly than armed forces can themselves.

This creates a number of challenges. The first is that civilian information infrastructures complicate the military's effort to retain information advantage on the battlefield. The second is that the armed forces must maintain records so that they can develop lessons from past actions. At the same time, these records must be capable of standing up to the kind of close scrutiny that emerges when soldiers are prosecuted for war crimes. Finally, civilian participants are capable of recording and uploading events, exposing the gaps between the claims armed forces make and the realities of what they are doing.

But the challenges of digitalisation do not end there, for it is also already clear that multiple data trajectories obscure and shape how battle is both perceived and fought.[6] And adversaries themselves are seeking to manipulate the new war ecology to spread uncertainty, sow confusion and create space for strategic advantage. The result is a strange juxtaposition where accelerating warfare with the application of algorithms and precision targeting dehumanises the act of war even as the enemy successfully humanises it in the online narratives they develop to motivate supporters and intimidate adversaries.

As we explain in what follows, this ruptures the experience of battle. One military event may produce multiple data trajectories. These trajectories do not spin off the battlefield at the same speed. The result manifests itself along two main dimensions. The first is an impulse to increase the tempo of military operations so that the armed forces stand a chance of moving as quickly as public informational trajectories. The second reveals that the governmental process of making sense of official or open source data appropriate for generating lessons from war cannot keep up with the speed at which kinetic operations themselves proceed. The net result leads to armed forces embracing innovations such as AI, roboticisation and autonomous systems that allow them to process datasets at the speed at which data moves without being able to institutionalise knowledge across the entire organisation.

All of this results in a disjuncture between the acceleration of warfare, the digitalisation of government practice and approaches to state record-keeping. This is further being exploited by technologists who create 'on the fly' archives. Unable to effectively manage or maintain these archives, the state must rely on private information infrastructures to disseminate knowledge. This creates a space for government officials to use WhatsApp to make decisions while at the same time asserting that these communications are not subject to scrutiny or freedom of information requests.[7] This in turn represents another hollowing out of the state with significant ramifications for the way we produce military history and memory in the new war ecology. However, to understand how archives, history and war are enfolded, we must start by looking back at the networked battlefield that emerged from wars in Iraq and Afghanistan as part of the GWOT.

Data Fusion and the GWOT Battlefield

Among those Western states 'allied or aligned to the United States' and who 'fight using methods pioneered and perfected by the United States', instant, remote war now dominates the military's way of imagining war (Kilcullen 2020). Although these ideas reflect twentieth-century cybernetic thinking, the capacity to fuse multiple public/private data sources in such a way as to facilitate the real-time precision targeting of individuals through to battlefield assets was only properly pioneered by the United States during the Iraq War. In support of this process, the American armed forces pioneered the development of intelligence Fusion Cells. These Fusion Cells brought together a range of intelligence agencies who could identify and reconstruct social networks, mining data captured by Special Forces and recycling this into follow-up raids that same night. The combination of systems and data integration has given the West a very powerful military lever that is now being used to manage threats that emerge from across the globe.

For those wishing to challenge the existing world order, however, neutralising Western advantage from positions of military weakness would appear to be a forlorn hope. And yet, as the GWOT battlefields demonstrate, geography, IEDs and the mobile phone have been effective counters to the evolution of Western military techniques (Kilcullen 2020). In what some might describe as the 'Shock of the Old', old technology has been repurposed to surprising military effect (Edgerton 2008). Jerry-built weapons using readily available software applications from mobile phone/tablet operating systems are put to service to help insurgents engage adversaries, using the advantages of open societies to increase weapon effectiveness (Hashim 2018). As a result, a forward observer using a mobile phone and the Google Earth app can communicate with a distant fireteam, thereby increasing the accurate targeting by analogue devices such as the mortar. By themselves, these technologies cannot produce victory, but used carefully they help to channel adversaries into cityscapes and restricted landscapes. This in turn makes it easier to spring ambushes and create tactical advantage.

Poor training and lack of tactical sophistication is a challenge for insurgents; however, the goal is not to go head-to-head in a firefight with professional armies but to gain maximum political leverage from every engagement (Hashim 2018). What is important is not the death count but creating a spectacle that can be transmitted over the web. In their planning, insurgents can work out the best location for filming an IED explosion. This can then be recorded and shared online, amplifying the message of resistance. Not only does this solidify support among those who might be persuaded to join an insurgency but it also demonstrates an understanding that they must weaponise the open societies of the West in an effort to undermine the counterinsurgents' will to fight. Thus, for those engaged in trying to resist the West's military dominance, digital spaces have created opportunities for amplifying the effects of war – neutralising and undermining the West's preference for transferring risk (Shaw 2005) – by weaponising the online world for the purpose of advancing political agendas.

Given the nature of the terrain and the structures of the societies in Iraq and Afghanistan, insurgents had to perfect a number of different tactics in order to defeat the West's intervention forces. It is difficult to generalise, but in relation to the war in Iraq, insurgents preferred to operate in cities, goading adversaries into overreacting and using suicide bombers to cause mass death and create insecurity between religious communities (Ford and Michaels 2011). By contrast, Afghanistan is a sparsely populated and predominantly rural environment, which demanded significant troop deployments if a permanent presence on the ground was to be created. This proved to be an impossible commitment for Western governments, as a result of which counterinsurgents in the middle years of the campaign found themselves attempting to sweep and clear insurgents even as the Taliban retreated to its sanctuaries in Pakistan (Chaudhuri and Farrell 2011). In both wars, the West struggled to contain and then dismantle the insurgent networks that sustained the operations against US and allied forces.

It is not easy to map and trace the emergence of an information environment that has made it possible to develop a military doctrine that denies the West's advantages in remote technologies, systems integration and precision munitions. What is clear though is that a combination of tools and devices have been crucial for shaping the new war ecology. These weaponise Western information infrastructures against the West's armed forces and in ways that produce a media environment that cannot be controlled by a combatant on their own. Crucial among these technologies is the emergence of Web 2.0 – an evolution of the web that allowed users to publish and collaborate as co-creators of user-generated content – and the widespread availability of smartphones. When combined, both of these systems made it easier to track and monetise public engagement by collecting data on who visited a site, what they looked at and what sites they went on to look at. Designed to enable users to more easily create their own web content, tag it, photograph and video it and then share it on social media, these systems have been embraced by users.

The digital world's growth from West to the rest has not occurred evenly. This has produced opportunities for those wanting to misdirect Western military technique and exploit battlefield events for political advantage. Given the reliance on digital media and the internet for amplifying the results of battle, an approach designed to generate confusion and misdirection relied on a great deal of media preparation of the battlefield. The challenge would be to create ratios of digital noise to actual signal that would slow down the ability of analysts to trace and identify fact from fiction. If secondary witnesses to an event might also be harnessed to further magnify a story and it was repeated often enough, then it might lead an event to make its way from the outer reaches of the internet and gain credibility as it found its way on to mainstream news reporting websites.

By way of contrast, while the West's way of thinking about war has been shaped by a determination to perfect remote, instant warfare, the hierarchical command structures that have traditionally made it possible for Western nations to produce military power have struggled with the management of twenty-first-century news media. These media management issues have their roots in two wars of the twentieth century. During the Vietnam War, American commanders took a relaxed approach to controlling the movement of journalists, allowing them to visit battlefields without military escort. This made it possible to gain a better idea about the ebb and flow of the counterinsurgency in Vietnam, but it also meant that US commanders could not control how American citizens might see the war. As a result, Vietnam produced a culture of suspicion between the military who believed journalists did not understand war and were undermining their efforts, and journalists suspicious of the military who were unsure whether US commanders were hiding something (Rid and Hecker 2009).

By the time of the 1991 Gulf War, however, news reporters found themselves under tight 'top-down' media management (Merrin 2018). In order to control journalists, a 'pool system' was created 'whereby a select number of predominantly Anglo-American journalists were accredited by the military and allowed to operate alongside troops' (Rid and Hecker 2009). Despite the technical capacity to broadcast near instantly from the battlefield to newsrooms twenty-four hours a day, seven days a week, the military and the news media experienced little friction while working together. This was partly because the war was over so quickly and partly because the destruction of Saddam Hussein's armed forces largely occurred over the horizon as a result of ordnance delivered by air.

In terms of martial culture, the American armed forces were ill-disposed to journalists, viewing the media suspiciously and as something to be managed even as the main effort involved delivering kinetic effects on to the battlefield. Consequently, during the 2003 Iraq War, when confronted by irregular adversaries willing to make use of Web 2.0 and smartphone technology for propaganda effect, US commanders initially struggled to adjust to the new media environment in the same way as conventional MSM struggled with citizen journalists. Bound by hierarchical approaches to media management, seeking permission from more senior levels in the command chain, the military took time to adjust to the new media conditions it faced in Iraq and Afghanistan (Rid 2007).

A conventional military hierarchy, media illiteracy and a certain wariness around the cyber-security implications associated with using web-based platforms left the Americans flatfooted when it came to winning the war for public opinion in online spaces. By contrast, jihadist media operations were very carefully managed so as to tightly relate online and offline propaganda to military activities. The goal was to unify friends, deceive enemies and prepare the information environment so they could convey their messages at a faster pace to their American opponents (Whiteside 2020). This conveyed an advantage over American media operations, helping to shape the tactical and operational environment through deception or misdirection. More than this, it could play out in beneficial ways for those insurgents who understood the importance of driving home their political points.

The speed and success of the insurgents in turn demanded more from American commanders, who had to script out the potential media war so that when it came to a planned combat engagement, the media messages were already approved by higher headquarters and prepared to be released online and in the MSM. The result was an evolution in military practice that placed influence and media operations alongside the use of kinetic effects for the purposes of achieving successful military and political outcomes. By the late 2000s, US and coalition military doctrine had evolved to the point where an operation might seek to deny an enemy access to terrain or a population group and at the same time try to develop a media message that would consolidate support among the convinced while trying to attract the support of the uncommitted.

The effectiveness of US information warfare doctrine was, nevertheless, open to considerable debate.[8] Indeed, given the complexity of the information environment that US forces were operating within and trying to shape – subordinate military commands, global news media, local news agencies, overseas and domestic press, citizen journalists, NGOs and local bloggers and vloggers – it is no surprise that the process of identifying target audiences and controlling how the message emerged was complicated and not always successful. When this was combined with the range of disparate organisations involved in identifying and shaping the information environment – from DoD public affairs and public diplomacy officers all the way to State Department, CIA and in theatre psychological warfare/information operations officers – the complexity of the US military meant that quick responses to enemy propaganda were sometimes hard to come by.

Even as Western forces developed techniques to counter the infowar, British and American forces attempted to iterate military practice so that they could operate more quickly than their adversaries. This led commanders to develop a military tool for defeating the adversary even before they had managed to get their information war off the ground. Crucial to this was the adoption of practices that mirrored those of the insurgents themselves. Famously declaring that 'it takes a network to defeat a network', General Stanley McChrystal, the 2003–8 commander of the Joint Special Operations Command, adopted an approach that involved fusing the analysis of different data sources into a Fusion Cell that could be used to direct military activity.[9] This emphasised the identification of the insurgent network followed by the targeting of key people or nodes within that network. The military doctrine that emerged out of this stressed the utility of lightning speed and successive, near instant multiple blows with a view to unbalancing the enemy so that they could not launch their own terror attacks. Culminating in a doctrine known as Find, Fix, Finish, Exploit, Analyse (F3EA), the enterprise depended on fusing information from multiple intelligence sources and agencies including for example mobile phone use, image intelligence and local informants (Ford 2012). This was then provided to Special Forces to capture or kill targets and gather up information that could be used to further identify and break down the insurgent social network. Operating at speed, Special Forces might go out several times a night to strike successive and multiple targets, using intelligence gained from one location to identify the next target.

By attacking the insurgents' network, the goal of these counter-terror raids was to defeat insurgent cells before they were ready to strike and so create a period of time for community leaders to negotiate and reach some sort of political settlement without fear of terror attack. If part of a network could be disrupted, then it would leave other terror cells in the organisation uncertain as to whether they might be struck next. This in turn created a number of further problems. In the first instance, insurgents were driven underground, breaking down their cell structure into more compartments so as to make it harder to successively attack each unit and roll up the network. In the second, this might make terrorists more inclined to engage in even more dramatic displays of violence with the aim of luring their enemies into using more force than was otherwise necessary. By broadcasting this escalatory dynamic, the insurgents could further discredit the counterinsurgents as predatory invaders (Urban 2010).

Unfortunately for the Americans, however, a strategy that aimed 'to buy time for the invariably slower improvements in governance to occur' rather than coming to a direct political settlement was not guaranteed to produce political success (Farrell 2017). In Afghanistan, for example, as one former Taliban government minister, and in 2011 a leading member of the insurgent's propaganda cell, observed: 'We never have calendars, watches, or calculators like the Americans do'; rather, '[f]rom the Taliban point of view, time has not even started yet'.[10] The American soldier, he said, 'starts his stopwatch, counting every second, minute, and hour until he gets home'. By contrast, '[o]ur young fighters ... are not thinking of time and consequences, only of the endless fight for victory. These fighters measure time by how long it takes to grow their hair.'[11]

The implicit contradiction contained in the speed of American operations was that these techniques could not produce a political victory against an adversary who was willing to take casualties and use this for online propaganda purposes. Nor could American operations defeat an adversary who could find sanctuary in countries that Western armed forces were unprepared to invade or who could hide in geographies that were difficult to control (Innes, 2021). All the Americans could do was throttle the speed at which an enemy might undertake their own offensive operations. Defeating the insurgency was beyond their capability.

That is not to say that image and surveillance intelligence when combined with data-mining tools could not be extremely successful at triangulating on terror cells that broke cover and used modern communication devices to organise their activities. However, the insurgents would not negotiate with those who denied them their legitimate political rights.[12] Consequently, despite its sophistication, the paradox of a military technique explicitly designed to attack insurgent networks was that in practice it involved keeping a lid on enemy activity rather than bringing them to accept defeat. Worse than this, even as the quagmire in Iraq and then Afghanistan unfolded, the nine-year search for Osama bin Laden implied that Western forces were impotent against an adversary who was willing to commit themselves wholeheartedly to their cause. Indeed, by staying off the phone and internet grid, bin Laden successfully demonstrated a way to hide in Pakistan, an American proxy state, and still coordinate a worldwide terror network (Owen 2013). This gave the Islamists ample time to keep their movement alive, take inspiration from bin Laden and organise their own form of political and military resistance, taking what had been learnt in Iraq and Afghanistan and honing it for future wars across the Levant (Hashim 2018). Far from producing the *fait accompli* inherent in the West's vision of the battlefield, insurgencies in Iraq and Afghanistan produced perpetual war. The West may have hoped to bring about quick victories, but repeated missteps helped their enemies gain control of the timetable.

The Paradox of Accelerating War

The battlefields of the GWOT have revealed the way that information infrastructures, data and military activity mesh, helping to incubate the new war ecology of the twenty-first century. Military activity inevitably moves at a different tempo from the data trajectories that it produces. This is a function of the new war ecology, where the military cannot control all the sources of data on the battlefield. Traditionally, a typical way to control the messaging around military activity has been to prepare the information environment with scripts in advance of events. This would involve releasing a pre-prepared news story to accompany a raid or an attack in the hope of managing the way an event was portrayed. Surprise and cyber-security could then be used by the armed forces to control the speed at which various narratives might emerge. What we are now seeing, however, is that these techniques also obscure and hide the way stories about events are told and in ways that are as helpful as they are a source of conspiracy theory. The Holy Grail, then, is to speed up military activity so that armed forces can keep ahead of and control the narratives that emerge out of the new war ecology even as they distract and try to slow down the counter-narratives of their adversaries. The danger is that the speed of military activity undermines political oversight and paradoxically renders war detached from its strategic rationale.

Within this frame of analysis, then, the West seeks technologies and doctrines that allow it to further accelerate the way that it can deliver kinetic effects to allow the military to stay ahead of the strategic narrative (Miskimmon, O'Loughlin and Roselle 2013) even as their activities continue to be directed by the state. To do this effectively, however, Western armed forces not only need to be nimbler in how they manipulate the new war ecology but must also be faster than their adversaries on the battlefield. This means speeding up processes of military innovation and moving towards what some describe as 'prototype' or 'beta' warfare.[13] Based on the idea that complete solutions can never be fielded into battle, the suggestion is that armed forces should prepare to work with the best prototype that they can field in order to gain a competitive edge. As the chief of the Australian Army argues, this would be considered an appropriate response because '[f]uture advantage will lie with the side who can "own the time" and best prepare the environment'.[14]

The underlying technological challenge posed by a conception of warfare that values owning time over other variables lies in organising the instruments of power to deliver military effect instantaneously. This has culminated in an increasing interest in AI, roboticisation and automation, where algorithms can quickly and more accurately process vast quantities of data than humans. This cybernetic approach to technology points to a unifying rationality that brings human and machine together as part of an idealised military project where targets are struck even before an enemy has an opportunity to act. As outlined in the military theory of Colonel John Boyd, an American fighter pilot during the Korean War, much of modern Western military thought is implicitly framed by the notion of defeating an enemy by making decisions more quickly than an adversary and thus getting inside the enemy's decision-making cycle (Osinga 2007). Born out of his experience of flying F-86 Sabres against MiG-15s, Boyd's theory was based on the hypothesis that intelligent organisms and organisations work through a continuous cycle of interaction with their environment, adjusting behaviours in light of the information gathered. Breaking down the process into four steps he called Observe, Orientate, Decide, Act (OODA), Boyd claimed that the successful pilot would go through the OODA loop process more quickly than the pilot who found themselves shot down. As Frans Osinga observes, the significance of the OODA loop lies not in the way pilots or commanders might choose to think about their approach to decision-making and defeating their enemies but in how the OODA loop concept can be applied to cybernetic systems. Thus, the notion of the OODA loop has been taken to mean a collapse in the human–machine binary with the effect that it has now systematised the way armed forces implicitly think about the development of autonomous weapon systems (Scharre 2018).

The OODA loop symbolises much of the technicism that lies at the heart of the Western approach to warfare. As Grégoire Chamayou has articulated, however, the current apotheosis of this line of thinking now seeks to 'eradicate all direct reciprocity in any exposure to hostile violence' (Chamayou 2015, p. 17). Wrapped up in a category of war that can best be described as remote warfare, this formulation of the utopian battlefield permits the West to prosecute violence without risk to anything other than expendable resources, whether they are proxy alliance partners, private security companies or remote technologies. Western war is thus a process of managing violence to the peripheries of the developed world, where a strategy of 'mowing the lawn' has more resonance than seeking decisive battles that produce victors and vanquished.[15]

Remote and autonomous systems are clearly designed to avoid reciprocal violence while offering armed forces the opportunity to deliver immediate military effect (Renic 2020). The goal is to produce a *fait accompli*. To avoid suffering some sort of setback, in its purest form the objective is to do this at great speed: to identify a target and strike it instantly before the enemy can decide to act differently. Virilio describes the philosophy that underpins these notions as '[o]nce we have seen something, we have already started to destroy it' (quoted in Bousquet 2018). To destroy something instantaneously is to accept that judgements about a target's value have already been reached and there is no further need to discuss or reconsider. Thus, the West's utopian battlefield is concerned with applying military violence instantly, at distance, without reciprocal threat and, once a decision to use force has been taken, without entering into dialogue with an adversary.

Instant, remote war is, therefore, risk transfer war at its purest, helping politicians minimise the chance of punishment at the polls while maximising potential gains (Shaw 2005). Moreover, reflecting the views of the cyberneticists of the 1950s and 1960s, by removing the human from the decision-making loop you accelerate war by overcoming the possibility of human error (Bousquet 2018). To its advocates, this is the utopian battlefield, where death is administered humanely. Once targets have been selected, they can be destroyed without fear of loss to one's own forces. Human exposure to violence is limited to those people who are to be killed. Warfare becomes surgery, the clinical application of the knife to excise adversaries. The warrior ethos in the human–machine dialectic is subordinated in favour of technologies that do not demand self-sacrifice or bravery. Delivering the *fait accompli* implicitly limits the possibility that an adversary can find an opportunity to express a different political perspective. To those facing it, this is not war as Clausewitz described it. There is no duel between adversaries. It is a manhunt more than it is a continuation of politics by other means. This is the war of the coward (Chamayou 2012, 2015).

When framed within a Clausewitzian ontology of war, however, the danger is that these automated military techniques find themselves unhinged from politics. This is even more the case given that the datafication saturates decision-making, and algorithms are written and overwritten in ways that are opaque to the engineers responsible for coding these systems (Lindsay 2020). In these circumstances, Clausewitz's dictum that war was a continuation of politics by other means would find itself overturned by technologies that guaranteed political violence would perpetuate itself along an automated trajectory such that war would become an end in itself. That is to say, where politics as a process of negotiation within government could not keep up with military decision-making there would be a fundamental risk to the stable balance of power (Horowitz 2019a, 2019b). The underlying reason for this, as Virilio observes, is that '[n]o politics is possible at the scale of the speed of light. Politics depends on having time for reflection' (Virilio 2002, p. 43). Consequently, accelerated warfare negates politics. Once the button is pressed, destruction follows. The more automation there is in place, the less opportunity there is to change military actions in order to realise further political engagement. The more that people are withdrawn from the decision-making loop, the more the chance that war becomes an end in itself.

All of this has not dissuaded those who are subject to these expressions of Western military power from developing counter-strategies that create doubt or keep conflict open-ended so as to manage the West's infatuation with instant, remote warfare. Indeed, as the wars in Iraq, Syria, Crimea and Donetsk reveal, the tempo of military power can be exploited by those online clickbaiters who now play an important role in shaping the new war ecology. IS, for example, used its social media engagement to portray savage acts of brutality against its adversaries and demonstrate the unity of support among those who were building the caliphate (Almohammad and Winter 2019). At the same time, Western governments could not prevent these narratives appearing and consequently struggled to restrict their citizens from intervening on behalf of the jihadist movement. Similarly, in Crimea and Donetsk we have seen Russians successfully confuse and manipulate Western audiences through disinformation, thereby creating doubt over whether and how to support Ukraine.[16] But what also makes this kind of war so difficult to resist is not just the online amplification and targeting of messages – it's 'that those messages are often unwittingly delivered not by trolls or bots, but by authentic local voices' (Jankowicz 2020, p. 3). Accelerated warfare may help to speed up and sequence the delivery of kinetic effects, but for those participants streaming attacks via social media, the collapse of categories like civilian, combatant and participant is enough to disrupt the emergence of top-down political narratives.

Thus, the move towards accelerated warfare reminds us that 'time is a political good that is used when states and political subjects transact over power' (Cohen 2018, p. 4). These transactions take time and are sometimes deliberately obtuse so that battlefield realities and political understanding can be brought into sync (Stoker 2019). In such circumstances, considered and reflective government works at the pace of politics itself, in the personal offices of the politicians and their respective networks of influence and not at the pace of instant, remote warfare. For good government is not always fast government. Sometimes good government, like the military histories that underpin the technologies of martial experience that make for good generals, also takes time and reflection. In this respect, the uneven distribution of the new war ecology within the military, across government and among civil society tells us something about the latency of data as it informs policy, drives tactical change and shapes history and memory.

Data Trajectories and the Uneven Decay of Historical Distance

Patterns of digitalisation accelerate political and military decision-making even as broadcast media have continued to play a part in shaping narratives for those audiences who remain wedded to twentieth-century forms of communication. At the same time, other sections of the public can engage with the atrocities of war with an imminence never previously experienced. And the rate at which these processes have unfolded reflects a disjuncture in the cultures of government, the broadcast era and contemporary participatory journalism that now forms part of the new war ecology. The result has seen multiple and distinct cycles of sense-making and remembering where different audiences in the armed forces, within governments and across society perceive war in disconnected temporalities. This disjuncture is a feature of the new war ecology that only serves to amplify opacity, uncertainty and undermine consensus and social cohesion.

Like other organisations affected by late twentieth-century trends in globalisation, outsourcing and process re-engineering (Turner 2008), digitisation has stimulated an enormous proliferation of records while divesting data ownership, control and cyber-security to organisations outside the armed forces and beyond their associated defence bureaucracy. This has seen a significant growth in the provision of security clearances to US government contractors and outsourced businesses while also making the extended data infrastructures vulnerable to cyber-attack.[17]

As data leaks and gets hacked, government seeks to secure information through greater compartmentalisation, making more use of closed computer networks and applying higher security classifications to material. While these restrictions secure data networks, they also have consequences for record-keeping, data searching and accessing sources of information that the government and its contractors have produced. This makes it harder for bureaucrats to establish what the organisation does and does not know and leaves open the possibility that open-source information can be weaponised more quickly than military bureaucracies can check records.

To try to get round these problems, government bureaucracies are seeking to leverage cloud-based data hosting. This ought to simplify the control of data, but it also further concentrates power in the hands of those companies capable of maintaining advanced information infrastructures. Thus, the UK security services contract Amazon Web Services to host their classified data in the cloud. Their hope is that this will afford an opportunity to make use of sophisticated data analytics and AI that will further their capacity to make sense of the source material they have collected.[18] Yet it also demonstrates that the state cannot speedily interpret the quantity of records it has created up to this point.

These efforts stand in contrast with how bureaucracies have managed their records in the first two decades of the twenty-first century. Up until this point, the cost savings associated with going 'paperless' have seen the military bureaucracy discard older, analogue practices of record management. These have not been replaced with appropriate data management methods suitable for the armed forces at war.[19] In any case, even if this happened in a way that made it possible to easily access future records, government bureaucracies would still be looking at a minimum twenty-year hole in their record-keeping. Consequently, as American Iraq War veteran Major John Spencer observes, the result is that

> for those wars with no living veterans – whether the American Revolution or World War I – we can remember. We can access digital archives of battlefield maps. We can examine lists online of personnel who fought in each battle. We can read orders from commanders, or personal diaries, journals and letters sent by soldiers to their loved ones. Unfortunately, our recent conflicts will be difficult to remember in this way.[20]

As Spencer continues, this has implications for how armed forces prepare for and think about future war.

When it comes to the military bureaucracy and its capacity to make sense of the battlefield, then, the cumulative result of twenty years of digitalisation in the twenty-first century has been uneven and sometimes paradoxical. Repeated bureaucratic reorganisation has complicated record-keeping.[21] This reflects the uneven distribution of information infrastructures across government and the armed forces and the distinct cultures of record-keeping that emerge from them. For example, when it comes to the British military there are at least two distinct cultures that shape organisational memory and strategy, reflecting the perspectives of the services themselves and that of the Ministry of Defence (MoD) more broadly. In the case of the MoD, the continuity and longevity of the archival record of military operations conducted by the services is ultimately the principal responsibility of the Civil Service. Here civil servants in the respective Historical Branches of the Army, RAF and Royal Navy must manage records in accordance with the stipulations of the Public Records Act of 1958, as subsequently heavily amended by the Freedom of Information Act 2005.

Since the start of the twenty-first century, however, the processes by which records are kept and managed have come under considerable strain through digitalisation, public inquiry and declining standards in archiving practices (Moss and Thomas 2017). The Army Historical Branch, for example, now collects terabytes of data. This relies on the Army retaining rather than repurposing computer hard drives and involves civil servants going to headquarters to collect material before it is destroyed. Even assuming this process happens unproblematically, the archive from operations in Iraq and

Afghanistan is bigger than anything previously created by the armed forces at war. Indeed, to make the comparison meaningful, while 1 terabyte might hold as many as 143 million pages of Microsoft Word documents, the entirety of the UK's National Archive catalogue searches just 32 million descriptions of records created by various branches of the UK government.[22]

Setting aside the volumes of data, the civil servants who manage this material are located at fixed MoD sites in the UK and work according to different legal and career parameters from those of the armed forces. This compares with the challenge facing the services, who maintain multiple military locations across the world and adopt a policy of limiting an individual's service time in a role or while on operations to a tour of duty. The notion of a 'two-year tour' was developed to maintain the effectiveness of a fighting force and sustain morale during times of peace. At the same time, by moving officers around the Army it became possible to build organisational resilience and provide an opportunity to gain the requisite experience for career progression. The flip side of this mode of using human resources has been a decline in the continuity of military record-keeping for an organisation whose records are typically created in different locations across the world. In effect, then, the MoD has many thousands of sites, significantly more than most UK government departments, and military personnel rotating in and out of these locations on regular cycles. Consequently, the cultures of record management differ depending on whether you're in uniform or the Civil Service.

Given the difficulties of keeping its records accessible and up to date, we see how the armed forces are now even more dependent on the new archives of war (Agostinho et al. 2021). These are based in the cloud, on social media platforms and generated by everyday participants whether in uniform or not. These new archives are open, networked, connected, mobile, always on and carried everywhere. They are constituted on-the-fly by people uploading experiences from their everyday lives to the web and so do not conform with the traditional practices that have shaped the creation of state archives. These new archives have been successful in framing military activity and reminding the armed forces that they do not control the representations of battle. This in turn reminds the armed forces of their reliance on twentieth-century approaches to record management and their patchy ability to save, store, access, mine and deploy the data that they themselves produce. This has important ramifications for framing both battlespace management and information warfare and, just as importantly, shaping public narratives and understanding of war.

As a result, we are now witnessing the rupturing of the state's capacity to come to a full understanding of the wars that it has started. This process is born out of trends in the uneven distribution of information infrastructures within government and can be most clearly seen in the way that the state collects battlefield data and uses it to produce quick reflection pieces or tries to defend itself when soldiers are accused of war crimes. In each of these uses of government-collected data, we have to ask what is being collected by whom and how it is used in such a way as to generate legitimacy within defence and wider civil society. In this respect, the notion of 'historical distance', as described by Mark Salber Phillips, is relevant and may manifest itself 'along a gradient of distances, including proximity or immediacy as well as remoteness or detachment' (Salber Phillips 2004, p. 89). This gradient of historical distance is critical for shaping public and political perceptions of the legitimacy or otherwise of inquiry work.

The success of inquiries to shape ongoing and future practices in policy and strategy thus depends on how data is distributed out from the battlefield. Data emerges in different parts of the bureaucracy at different times and in ways that reflect operational priorities and the vagaries of compartmentalised approaches to official secrets. The sheer quantity of documents represents a gold mine for future military historians. However, that very quantity of material betrays the serious problems facing those members of the armed forces seeking to generate insights from previous experience, a challenge that is typically embedded in the opaque phrase, commonly found in military discourse, of 'learning lessons'. The US military withdrawal from Afghanistan on 30 August 2021 produced an avalanche of media stories proclaiming that 'lessons must be learned' after the failure of twenty years of war. What was not made clear, however, was how those lessons could be learned, by whom and to what ends? As Dan Spokojny observes, '[t]hese lessons are empty wishes without national security institutions capable of actually learning and evolving'.[23] Thus, the exact process by which organisational learning takes place, a process that is such a central feature of military progress and a critical component of inquiry work, is actually often taken for granted, even as the reasons for organisational forgetting are rarely made explicit.

Learning Lessons from Torture and Bombs

One way to resolve the time-limited and sometimes obscure workings of contemporary organisational memory and facilitate the learning of military lessons in an institutionally sanctioned form is through the production of Post-Operation Reports (PORs), campaign histories and Official Histories (OHs). Here we see how data trajectories interact with the inner workings of the military bureaucracy. Working to distinct temporal cycles, PORs allow military formations to quickly effect change within a campaign. Campaign histories might be used to generate lessons from a particular operation and longer-term OHs might be used by a Department of State to come to a formal and departmentally agreed position on its role in a particular set of events. Whereas PORs do not require much more than a recognition that new tactics, techniques and procedures might generate success, OH might take considerable time as they represent a formal acknowledgement of the need to learn organisation-wide lessons and implement institutional change. Thus, OH demand a greater level of scrutiny, objectivity and balance when compared with those modes of organisational learning that work to faster cycle times and are typically associated with the academic literature on military adaptation, innovation and effectiveness (see, for example, Farrell 2017; Kollars 2014; Catignani 2012; Russell 2010).

There are countless examples of these types of learning activities over the last twenty years. For instance, the US Special Inspector General for Afghanistan Reconstruction (SIGAR) set up an $11 million 'Lessons Learned' programme to diagnose policy failures in Afghanistan. Leading to the production of a number of publicly available reports, these lessons learnt were partly constructed from interviews of over 600 people. However, when journalists from *The Washington Post* decided to write a story on SIGAR they found that the State Department, the Defense Department and the Drug Enforcement Agency had all moved to classify documents and restrict their release to the press.[24] That so many of the transcripts from these interviews were subsequently classified can partly be explained by the interviews having been conducted 'off the record, not for attribution' so that interviewees were 'especially candid and willing to explore their own failures and those of others'.[25] Nevertheless, the classification of documents also points to the way the military seek to moderate the public record and, whether intentionally or not, undermine the legitimacy of their own lessons learnt programmes.

Brigadier Ben Barry was responsible for writing the British Army's campaign history of Iraq in which he noted that few members of the British armed forces had 'a genuine understanding of the full ebb and flow of the land campaign during this period'.[26] By the time the MoD released the report following a Freedom of Information request in 2016, the Army could observe that the lessons had 'been learned in whole or in part'.

The idea that lessons have been learnt in whole or in part is nonetheless a conclusion that hides a variety of questions that we might otherwise ask about the military's approach to explaining its performance. Consider the insights that emerged out of the seven-year Iraq War Inquiry led by Sir John Chilcot (Hoskins and Ford 2017) and what this tells us about the MoD's approach to data capture and storage when compared with the quick manipulation of the everyday news agenda through social media. The inquiry itself lasted from November 2009 to February 2011, but it took a further five years before the report was published, a pause that was used to allow all those cited in evidence the opportunity to comment on the findings. Meanwhile of course, in Iraq and Afghanistan, we have seen insurgents use digital media to quickly undermine Western narratives that portray coalition forces as benign, a task that can only be made easier by the length of time it took the Chilcot Inquiry to consider the evidence, reflect on its implications and complete a carefully worded final report.

What the Chilcot Inquiry reveals is how a lack of resources dedicated to maintaining explicit and tacit knowledge within the MoD hampered the coordination of organisational memory within the British Army (Moss and Thomas 2017). This was particularly evident in the Army's approach to IEDs and the use of torture, and was a function of distinct civilian and military cultures in relation to record-keeping while at war. For example, the usual challenges to organisational memory were further constrained by the short tour duties of some civilians, as summarised in the conclusions to the Chilcot Report:

> The difficult working conditions for civilians in Iraq were reflected in short tour lengths and frequent leave breaks. Different departments adopted different arrangements throughout the Iraq campaign, leading to concerns about breaks in continuity, loss of momentum, lack of institutional memory and insufficient local knowledge.[27]

More than this, in evidence given to the inquiry on 21 July 2010, Lieutenant-General Sir Alistair Irwin (adjutant general from 2003 to 2005) underlined the challenge of deployments in the rapid dissipation of Army organisational memory:

> [I]n respect of an institution, the only lessons that are learned and put into effect are the ones that are put into effect immediately, because the nature of an institution, with the individuals in it passing in and out and changing jobs and so on, is that unless the lesson is applied immediately, it will never be remembered. That's one of the real difficulties about lessons learned.[28]

By comparison, lengthy periods of time focused on one type of war require significant organisational effort to unlearn what has happened before and relearn how to fight in a different way. Thus, the shift to counterinsurgency in Iraq and Afghanistan required a completely different approach to fighting compared with the preparations that had been made to engage in conventional combat during the 1991 Gulf War. Despite the British having had more, and more recent, experience than their American counterparts in fighting insurgency-style warfare in Northern Ireland, the Chilcot Report clearly recognised that the Army's organisational memory had faded quickly. Indeed, as Lieutenant-General Jonathon Riley described in his evidence to the Iraq Inquiry on 14 December 2009:

> [I]t was borne in on me very strongly how much the collective experience of the army of dealing with the IED threat had wasted out during the long period of ceasefire in Northern Ireland. We had forgotten institutionally how to deal with this ... not just as a series of devices but as a system and how to attack the device and attack the system behind it.[29]

Even more corrosively, however, it is in relation to the Army's failure to institutionalise decisions and legal practices associated with torture where we can really begin to see that a focus on learning lessons misses wider questions associated with organisational forgetting. This became very obvious after the publication of the Gage Inquiry into the death of Baha Mousa.[30] Held in custody by the Army in Basra in 2003, Mousa had been subject to intolerable treatment and illegal torture and had subsequently died. In his report into the incident, Sir William Gage noted that the British Army had been banned from using five techniques (hooding, white noise, sleep deprivation, food deprivation and painful stress positions) on prisoners since the Parker Inquiry of 1972. In the absence of any significant means for sustaining that knowledge within the MoD, the notion that these techniques had been prohibited had largely been lost. Consequently, at the time of the Iraq War there was 'no written policy or doctrine banning the practices'.[31] Inevitably, some scholars have concluded that the policy on the five torture techniques hadn't so much been lost as that the process of organisational forgetting had purposefully been orchestrated so that the Army might continue to practise what it had been told it could not (Bennett 2011). Whatever the truth of the matter, it is fair to conclude that before Baha Mousa, the MoD's lackadaisical approach to organisational forgetting represented something of a 'corporate failure'.[32]

Further organisational memory challenges were faced by the 2009–14 Al-Sweady Public Inquiry, which dismissed allegations that UK soldiers had mistreated and unlawfully killed Iraqis in 2004.[33] The MoD were unable to fulfil all the inquiry's disclosure requests. They also struggled to establish which documents and records existed that were pertinent to its remit. This in turn led the MoD to reflect upon whether the search of some extensive and complex record sets constituted a proportionate use of its resources.[34]

In light of the experience and reports of these public inquiries, the MoD's approach to organisational memory has had to improve. The conscientious effort to put in place the ability to mobilise the Army's organisational memory by taking advantage of the Army Historical Branch's resources and expertise has been critical in this respect.[35] However, the complexity and scale of digital documents and records that need to be maintained continue to pose huge challenges to military organisational memory. The culture of record-keeping has had to adapt from printing off material from electronic sources, so that the hardcopies might be incorporated into hardcopy filing, to one where more care is taken to avoid erroneously cleansing data and redeploying hard drives for different uses. By necessity, this has demanded the introduction of an electronic archive capability so that crucial data might be copied across before hard drives are repurposed.[36] Nonetheless, if robust processes are to be put in place that consciously balance decisions favouring organisational forgetting with those promoting organisational memory then these activities need to be sustained for emergent and ongoing campaigns. If managed carefully, not only would this help to maintain the Army's internal, political and public legitimacy but it would also help the MoD deal with various inquiries into the conduct of Britain's armed forces in times of war. As accusations of covering up war crimes in Afghanistan and Iraq continue to surface, it is clear that the MoD will have to properly sustain this effort, something that may be especially problematic for operations involving Special Forces.[37]

These examples reveal that the military bureaucratic capacity to learn lessons is dependent on the trajectory and decay of data as it traverses through differing informational infrastructures and the cultures and processes that make them up. Some of this is to do with the historical distance associated with a particular activity, where officers and personnel still have a career profile that depends on avoiding acrimony or difficult questions. At the same time, it is also the case that the military organisation sometimes prefers to ignore or forget rather than institutionalise a particularly contentious way of working. Alternatively, where a process relies on uncodified, tacit knowledge, the replacement of personnel can lead to accidental forgetting (MacKenzie and Spinardi 1995) where we see data leak out from what appears to be settled processes of record management.

Deriving Meaning from the Ruptured Battlefield

Identifying military weaknesses and optimising performance through engagement in history depends on information management and record-keeping that gathers up and codifies information and knowledge that is explicitly and tacitly embedded in people, processes and records. However, the challenges posed by the digitalisation of headquarters and government departments increasingly leave the armed forces reliant on the production of partial histories based on selected readings of the public record combined with interviews of officers who understand that these exercises have both career advancing and limiting effects. In these circumstances, then, the West's preference for a depoliticised and an objective campaign or military history (putting aside whether that was ever actually possible) inevitably succumbs to the structural challenges posed by digitalisation and the proxy histories it leads to.

The development of technologies that accelerate data fusion and promote instant, remote warfare have a dynamic that is at odds with the efforts of those trying to garner insights from past action based on the erratic distribution of data from the battlefield. This is a feature of contemporary information infrastructures that can only serve to make it harder for governments to understand why they have done what they have done. This reflects a rupturing with traditional record-keeping that betrays the deeply mediatised nature of twenty-first-century life. It is not just the case that the cloud-based web platforms that make up contemporary archives have their own user cultures and norms of behaviour, reproducing the further segmentation of audience and market. Nor is it just that the archives themselves are weaponised for the furthering of particular agendas. Rather it is that the platforms themselves create audiences by inscribing the values of their founders into the experiences of those who use them (Srnicek 2017). That the state cannot keep up with this approach to archiving partly reflects the limited availability of technologists with the appropriate skillsets but also points to a very real concentration of power in the hands of those organisations that now dominate cloud-based computing.[38] And all of this is bound to drive differing patterns of sense-making across the new war ecology.

The meaning of battle was once the preserve of soldiers as flesh-witnesses (Harari 2008), civil–military officials and the expert historians who had the time to access the records kept by the state. But this process has now broken down, fundamentally rewriting who is involved in deriving meaning from battle. This development starts with participative war (Merrin 2018 – see Appendix) but is made possible by processes of deep mediatisation that have 'radically changed the fields of perception' (Virilio 1989, p. 7). The upshot of this is that the battlefield, like other areas of online life, becomes a prism for multiple audiences to read what they want from their representations. Consequently, battle is reinforced as a site of multiple meanings where expert communities cannot determine interpretations by dint of access to the infrastructures of their recording. Instead, we see the collapse of the participant, the audience and the expert into one register in which all become weapons for advancing particular perspectives of war.

In this respect, then, the forlorn hope of the official historian is that the archival infrastructures of the future will facilitate opportunities for writing complete histories of battlefield events. Given the evolution of the new war ecology and the accelerated collapse of binary categories into everywhere war, the very possibility that historians will have sufficient historical distance to produce some semblance of an objective, depoliticised account of the past has finally been rendered inconceivable. Indeed, the notion of accelerated warfare further destabilises older categories of war and peace, leading to a situation in which the management of conflict supplants war to become a permanent feature of life.

The consequence of this process of digitalisation and datafication has ramifications for historians of war and conflict who will increasingly find their disciplines reframed by the digital archives of the twenty-first century. As one member of the UK military told us, 'the post-operational report is dead'.[39] In other words, historians need to radically reorient themselves to the form, volume and complexity of official digital records and the associated challenges of access, search and analysis.[40]

In these circumstances, historians of war may continue to find an archival home in the pre-digital era. Those historians working on more contemporary matters, however, will have to narrow down their field of perception and accept that their analyses are simply concerned with telling stories about war's representation. But these questions are nothing compared to the deep problems facing government. For an unwillingness to deal with the challenge of digitalisation and the archive has the potential to delegitimise official histories of war and, if war cannot be properly justified in history, to undermine the legitimacy of government itself.

PART 2

ATTENTION

Diagram 4: Attention

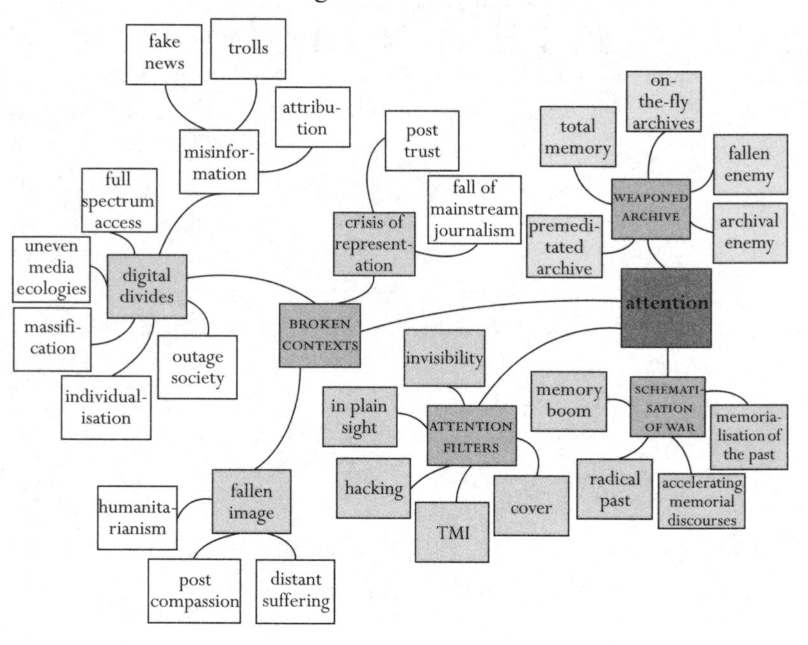

4

THE RADICAL PAST

Social media is a hotbed of falsehood and fact-free interaction. Indeed, one study found that a false tweet on Twitter reaches 1,500 people six times faster than an equivalent factual tweet.[1] Sometimes there is a relationship between a fact and a tweet's amplification. For the most part, though, Twitter users cannot be sure whether what is being tweeted has any basis in truth. Thus, the gap between a tweet's credibility and its virality opens up a space that only serves to feed doubt. This presents both opportunities and threats for those trying to influence how people think and understand the world. On the one hand, doubt can be used to manipulate perspectives. On the other, the different speeds at which true and false information moves around the web makes it possible for whole communities to get locked into a hall of mirrors where all that counts is the credibility of the moment.

With the emergence of the new war ecology, the archives of war have become less stable and more subject to the vagaries of digitalisation. This makes it considerably harder to break out of the hall of mirrors produced by social media. This either leads historians to complain that history is being debased or to find themselves studying what people remember rather than a contemporaneously written document. Consequently, our shared understanding of the past is caught between a pre-digital and highly sedimented appreciation for war in history as framed by analogue archives versus the digital churn of a present framed by social media. This agitation of history and memory is the radical past. If we are to make sense of this then we must fall back on mental shortcuts, or schemata, that can be invoked to help us organise, interpret and assimilate new experiences.

For scholars trying to understand the history and memory of war in the new war ecology, navigating this prism of mental shortcuts is a particular challenge. This is because datafication has transformed twenty-first century wars into an experience that is continuously being streamed. One way of creating sense out of this digital flux is to anchor war to history. The challenge for historians is that the historical method – investigating the past in depth, breadth and context – is not a field that is designed to work at a speed where all that counts is the credibility of the moment.

Historians may struggle to keep up with the speed of disinformation; however, this does not mean that people cannot find their own ways to make sense out of war. On the contrary, it is apparent that political violence in the twenty-first century is framed by older, more socially widespread memories about the relationship between war and society. This schematises understanding and acts as a lens through which online and offline representations of war are seen. The net effect creates a turning over of the past in which older representations of war get stacked on top of new discourses from the digital present. Thus, using the UK as an example, clichés about 'the war' or the 'Dunkirk spirit', appeasement or the empire become referents for explaining or drawing parallels with the contemporary conflicts that make their way across our social media feeds. By mapping out how these older and digital discourses come into contact with each other, this chapter explores the way different schemata of war in the twenty-first century structure attention.

In the twentieth century, memory has been framed by analogue media recorded on paper, magnetic tape and film. This form of memory now reflects a comparatively sedimented understanding of the past, where, for example, there is in many (although not in all) countries an established and routinised commemorative culture of the First and Second World Wars framed by, for example, Remembrance Sunday or Memorial Day. Memory in the context of Radical War, by contrast, is one that is driven by the immediacy and the uncertainty of digital infrastructures of recording, sharing and archiving, combined with an undermining of the traditional functions afforded to remembering that involve distance, reflection and hindsight. That is not to say that mainstream twentieth-century memory as represented in commemoration and history no longer has relevance. Rather we argue that this mainstream memorialisation of war does considerable work in framing how twenty-first-century wars are seen, legitimised or ignored.

The persistence of knowledge about past human suffering, particularly from war and genocide, has a social function in potentially diminishing the likelihood of such acts being repeated. This effort to sustain the lessons that prevent war from recurring leads society to constantly revisit its past through acts of commemoration. This represents both a form of moral obligation and a social performance and leads to what David Blight would describe as a 'rage for memorialisation' (Blight 2012). At the same time, the twentieth century's era of mass warfare has ushered in a kind of moral contract of mass remembrance and identity genuflection best embodied in the early twenty-first century in the form of First World War commemorations. This has fuelled a series of 'memory booms' where society has turned to commemoration in the belief that remembering the experiences of war will help guard against war's future recurrence (Huyssen 2000; see also Winter 1995, 2006; Sturken 1997; Wood 1999; Wertsch 2002; Müller 2002; Simpson 2006).

Radical War is shaped through an acceleration of these memorial discourses. The result of individual and collective memory working in online and offline contexts has splintered national narratives about the place of war in society. This has the effect of reframing how society makes sense of war, a process that is further complicated as transnational and global perspectives of war and remembrance are introduced. The result disorders attention and has contributed to a politics of polarisation, division and exclusion that has further politicised the history of war. Societies have always recast their pasts in light of present needs. The challenge facing societies in the twenty-first century, however, is that this process is no longer under the unique control of establishment figures in the nation-state. Instead, the commemoration of war is open to wider global discourses that challenge received interpretations of the past. As a result, and as Pomerantsev might describe it, we argue that the history of war as a discipline is weakened in this digitally aided post-truth or post-trust era in which 'nothing is true and everything is possible' (Pomerantsev 2015).

What is the Radical Past?

Memory is more elastic than history in that it is not something stable or settled. Rather, it is of the moment, an attitude and response towards a representation of the past that occurs in the present. Every time memory is remade it becomes active: a reactivated site of consciousness such that '[i]t is an imaginative reconstruction, or construction, built out of the relation of our attitude towards a whole active mass of organised past reactions or experience' (Bartlett 1932, p. 213; Brown and Hoskins 2010). Remembering is, then, dynamic, imaginative and counter-intuitively, an essential aspect of what it means to live in the present. As such, memory is always new and continually emergent, shaped by what is going on around us.

The problem with speaking of memory, however, is that it immediately dissolves into a myriad of modifiers. Indeed, some argue that the singular noun of memory makes little sense without such a modifier. Consequently, memory is always personal, cultural or collective (Roediger and Wertsch 2008, p. 10). Others prefer instead to use alternative terms, particularly in relation to the memory of war. American historian Jay Winter, for example, prefers the term 'remembrance' to memory, while the philosopher Eelco Runia argues there are 'two diverging approaches to the past, but instead of "history" and "memory", the poles of this opposition should be called "history" and "commemoration"' (Runia 2014, p. 4).

Whether memory or remembrance, if the impulse is to commemorate then David Lowenthal suggests a vital guide to the current oppositional state of public attitudes to the past. The first act is turning to the past as a refuge, 'as an antidote to present disappointments and future fears' (Lowenthal 2012, p. 2). The second is a perpetual living in the present in which '[t]he past has ever-diminishing salience for lives driven by today's feverish demands and delights'. This is a function of 'the sensory-laden penchant for computer gaming, coupled with attention-deficit hyperactivity disorder, betoken a here-and-now environment dominated by raw sensations' (Lowenthal 2012, p. 2). Both of these attitudes have in common a loss of attention through overload.

This pattern of refuge and distraction can be seen in the number of commemorations for twentieth-century conflicts. For Winter, the trajectory of this process of remembrance constitutes a series of memory booms (Winter 2006), the most vivid example of which can be seen in relation to the memorialisation of the Great War. This war was initially marked by periods of limited and mostly private recollection, denial, unspoken trauma and non-memory. Paul Connerton, for example, states:

> All sorts of institutional provisions were put in place to keep those mutilated soldiers out of public sight. Every year, the war dead were ceremonially remembered and the words 'lest we forget' ritually intoned; but these words, uttered in a pitch of ecclesiastical solemnity, referred to those who were now safely dead. The words did not refer to the survivors. The sight of them was discomforting, even shameful. They were like ghosts haunting the conscience of Europe. The living did not want to remember them; they wanted to forget them. (Connerton 2008, p. 69)

After the initial period of grief, it was subsequently commemorated in broadcast media and through public ceremony. The finale was the centenary commemorations between 2014 and 2018, which involved a number of governments staging ceremonies and events to mark the dates of battles as part of a series of public events.

In addition to the Great War, over the twentieth century the memorialisation of war evolved through cycles of commemoration. They did not move at the same speed. Winter, for example, identifies the time lag between mid- to late twentieth-century war and genocide and what he describes as the second memory boom from the 1970s. This reflects a balance between 'the creation, adaptation, and circulation' of memory, including 'when the victims of the Holocaust came out of the shadows' and when a wide public was 'finally, belatedly prepared to see them, honour them, and hear what they had to say' (Winter 2006, pp. 26–7). In the decade leading up to the centenary of the Great War, the relationship between those willing to discuss the past and society's appetite to listen had changed once again. Instead of there being a balance between the production and consumption of memory, the cycle of commemoration had accelerated into a new fervour for remembrance that appeared determined to make up for lost time.

Other wars have followed similar trajectories from memorial denial through to overexposure. For example, upon returning from Vietnam, injured and traumatised veterans were an uncomfortable and highly visible reminder of the war. Not only did they remind Americans that they had lost the war but they also embodied America's broken military and political status in the eyes of the world. As Connerton argues in his influential typology of seven types of forgetting, 'few things are more eloquent than a massive silence. And in the collusive silence brought on by a particular kind of collective shame there is detectable both a desire to forget and sometimes the actual effect of forgetting' (Connerton 2008, p. 67). This has taken on public form through memorialisation (Allison 2019) and led others to dispute the idea that veterans were ostracised upon their return from Vietnam (Lembcke 2000). But the myths of societal conflict post-Vietnam continue to shape contemporary public attitudes. Thus, as Ted Kotcheff from the 2017 Andrea Luka Zimmerman documentary *Erase and Forget* describes it, the treatment of some of the homecoming Vietnam veterans was such that '[t]hey were greeted with protestors holding signs up: "Baby Killers!" They threw garbage and human excrement over them, dead rabbits, baby dolls with blood all over them. Can you imagine being welcomed home like that?'[2]

A further example of a twentieth-century conflict with a long memorial gap was the 1982 Falklands War. This was a long-standing diplomatic argument between Argentina and the UK that bubbled over into a war for the South Atlantic islands. The UK's victory was followed by instant euphoria sufficient to sweep Prime Minister Margaret Thatcher back into power at the 1983 General Election. Even though the war had been successful, it did not appear to resonate in public memorial consciousness over successive anniversaries. Instead, the focus of attention was on el-Alamein and Singapore. Consequently, it was not until the twenty-fifth anniversary in 2007 that an official (state/military/media) commemoration of the end of the war was marked in the UK on a significant scale. What is notable here is that this was at a time when Winter's third memory boom was underway.

The relative collective silence in the UK over the Falklands' victory might be put down to any number of considerations. The tenth anniversary, for example, was masked by the difficulties faced by the Tory Party as they sought to win a general election while avoiding discussion of the party's 1990 coup against Margaret Thatcher. Facing the prospect of electoral defeat, John Major, the new prime minister, wanted to avoid commemorating the Falklands War and reminding voters of Thatcher's links to a number of political controversies associated with the war. Of these, the most politically divisive and the one that was picked up by many MPs in the Labour Party was the Royal Navy's sinking of the Argentine light cruiser the *General Belgrano* on 2 May 1982. This attack resulted in the loss of 323 lives, accounting for just under half of the Argentine deaths during the war. Subsequent questions about the incident centred on the legitimacy of sinking a ship whose proximity and direction of travel relative to the British-declared 200-mile Total Exclusion Zone around the Falkland Islands was contested. This controversy had not subsided twenty-five years later when the government announced a 'major celebration' of the Falklands War. Indeed, Tam Dalyell – a former Labour MP who had campaigned against Thatcher at the time of the sinking – said that any celebration would be a 'reckless, stupid, thing to do'.[3]

Events, competing commemorations and political controversies may have meant delays in marking the Falklands War, but the twenty-fifth anniversary offered the first moment of sufficient historical distance for the UK to officially commemorate the anniversary. This was nothing, however, compared to the thirtieth commemorations, which were an even bigger affair. The 2012 anniversary was heavily inflected with the politics of the present. This included the Tory Party's determination to honour Thatcher before she passed away and renewed rhetoric from Argentina as they restated their claim of sovereignty over the Malvinas. Given the way that online debates jarred with official representations of the war, these commemorations were more complex and critical than those from just five years earlier. For critics on the left, the Falklands commemoration was a nod to Britain's last imperial war. For those on the right, however, the thirty-year anniversary was an opportunity to celebrate the future Global Britain's place as an ongoing force for good.

The thread linking all three of these examples is that their memorialisation has been uneven, contested, silenced and modulated by a variety of personal and public traumas and the political demands of the day. As such, these wars have all traversed the arc of a memory boom to reach a point of established historical consciousness even as they are now becoming overexposed in a digital present. Indeed, as we discussed in Chapter 3, when it comes to earlier wars John Spencer notes that '[w]e can access digital archives of battlefield maps. We can examine lists online of personnel who fought in each battle. We can read orders from commanders, or personal diaries, journals and letters sent by soldiers to their loved ones.'[4] As these things are online already, it then becomes possible to link, discuss and return to these materials, constantly accumulating commentary on wars in ways that grab and saturate attention.

One reason for this relentless churning of attention is that the memorialisation of war is as much a process of creating emotional distance as it is an act of remembering the fallen. In this respect, commemoration is not really a matter of memory but is also designed to help society come to terms with and renegotiate how it engages with its past. In this context, one counter-intuitive strategy to deal with an uncomfortable or difficult past is to commemorate it to oblivion through 'the spectacular promotion of a phenomenon' (Baudrillard 1994, p. 23). As Baudrillard argues:

> Our societies have all become revisionistic: they are quietly rethinking everything, laundering their political crimes, their scandals, licking their wounds, fuelling their ends. Celebration and commemoration are themselves merely the soft form of necrophagous cannibalism, the homeopathic form of murder by easy stages. This is the work of the heirs, whose *ressentiment* towards the deceased is boundless. Museums, jubilees, festivals, complete works, the publication of the tiniest of unpublished fragments – all this shows that we are entering an active age of *ressentiment* and repentance. (1994, p. 22)

In a similar light, Runia states that '[t]he more we commemorate what we did, the more we transform ourselves into people who did not do it' (Runia 2014, p. 9). In this way, we can see commemoration as a way of managing blockage, as an effective strategy of forgetting.

However, in the context of the digital present, it is unclear how Western societies are getting past the blockage produced out of commemoration. This is partly a function of the widespread availability of so many different media channels but also the stacking of different commemorative moments on top of one another. Take, for example, the seventieth anniversary of Operation Dynamo, where the Royal Navy organised the sailing of little boats across the channel to rescue British and French troops stranded at Dunkirk. On the same day, 25 June 2009, Pink Floyd marked the thirtieth anniversary of their successful album, *The Wall*, and Michael Jackson fans commemorated the great singer's death. All these stories jostled for news time and attention, in the process affording them all with an informational equivalence in an implosion of marking dates. This collapse has been enabled by the web, which now forms a single archive, but it points to the way that commemoration has a particular role in helping the public identify a shared past. Indeed, as far as Runia is concerned, '[t]his desire to commemorate is, in my opinion, the prime historical phenomenon of our time' (Runia 2014, p. 2).

Commemoration, then, is a constituent part of and plays a defining role in the public narrative about war. This is the case for politically motivated commemoration where a specific date is used to frame a public narrative about the place of war in society. At the same time, commemoration can also form part of a narrative that a community of interested participants have chosen to mark. All of this fits into the cycle of Winter's second and third memory booms, the outward expression of which can be seen, for example, in the emergence of digital television channels like the British channels 'Yesterday' and 'Blighty'. Recycling old documentaries and soap operas taken from the TV archives, these channels enable viewers to indulge both in nostalgia and what Paul Gilroy calls a 'post-colonial melancholia' (Gilroy 2006).

The inability to break past these framing devices has led some satirists to make fun of the commemoration of everything. In 2007, for example, the BBC's *Broken News* satire – a parody on the flux of frenetic 'prediction, speculation and recap' that today passes as news – ran a spoof commemoration of 'Half-Way There Day', or the 'day that marked the halfway point between each end of World War II'. Commemoration may have been satirised, but the commemorative blockage could not be dislodged through comedy. In these circumstances, it seems that an excess of remembrance has led to a devaluation of what it means to commemorate. Consequently, the defining features of the current memory boom are that everything is memorialised, but nothing stands out in memory.

When looked at this way, the radical past is a constituency that is rooted in a period of rapidly changing technology. At its most simple, the radical past is analogue, set in marble, from a time before digital media. In its more complex form, it is the annual commemoration interlaced with special anniversaries that are themselves pockmarked by less popular memories. The existence of the radical past was an easily defined and a central feature of the pre-2000 paradigm of national remembrance. Harder to define and rationalise is the digital processes that are rewriting history daily in a relentless turning over of the past. This problem is the subject of the next section.

Memorialisation in the Digital Present

Memorialisation in the digital present occurs along a number of dimensions. In the twenty-first century, information about war is rapidly made available to people beyond a conflict zone, creating opportunities for heightened reflexivity and immediate remembrance. This in turn has facilitated a process of online commemoration as part of a virtual memorial. These new digital environments offer more open discussion and greater opportunities to contest the meaning of what is being commemorated (see also Knudsen and Stage's 2013 analysis of YouTube video tributes to fallen Danish soldiers in Afghanistan and Iraq). At the same time, the memorialisation of the digital present re-structures how commemoration is experienced, extending the marking of an event along a number of trajectories beyond the official public or national process of bereavement. In particular, these online moments do not produce a collective consciousness that has been reflected upon and had the benefit of some historical distance from the events in question. Instead, the connective turn ushers in a new mediatised memorial of events that runs alongside and goes on to reframe processes of national commemoration. Consequently, the radical past reshapes the bounds of national identity politics.

An example is the artist Joseph DeLappe's academic direction of the Iraqimemorial.org. This online project commemorates civilian deaths since the beginning of the Iraq War in March 2003 (cf. Hoskins and O'Loughlin 2010; Hoskins and Holdsworth 2015). DeLappe's aim is to bring memorialisation into sync with the continuing and continuous civilian deaths in the bloody aftermath of the Iraq War and to 'mobilise an international community of artists to contribute proposals that will represent a collective expression of memory, unity and peace' to 'create a context for the initiation of a process of symbolic, creative atonement'.[5] This work effectively premediates official national memorialisation through its call for 'proposal concepts' to memorialise civilian casualties in the Iraq War, with over 150 artists' works listed under the 'Exhibition of Memorial Concepts'. The site includes diagrams, plans for galleries, photographs, videos and mixed media exhibits and is also open to public views and ratings of entries in addition to those made by 'internationally based curators and scholars'. In effect, then, this memorial platform preconfigures national debates about how to commemorate the Iraq War. In terms of immediacy and continuousness as well as in the exhibits' sometimes hypothetical character, it also stands in contrast with the often slower commissioning and construction of more traditional memorial practices.

The relentless memorialisation of the digital present spills over into non-digital contexts. This is particularly noticeable in the interventions by artists who chart how unfolding forms of warfare might be understood. A notable case of this premediation of memorialisation and of the challenges it makes to existing forms and mechanisms of memory is the 'Memorial to the Iraq War' exhibition. Organised by the Institute of Contemporary Arts (ICA) in London in the spring of 2007, the memorial aimed to explore the potential alternatives for memorialisation of the Iraq War. As part of this exercise, twenty-six artists from Europe, America and the Middle East accepted an invitation to imagine a memorial, some of whom went on to present their work in the Institute's exhibition space.

One of the most notable exhibits was by Christoph Büchel, who drew parallels between a drug administration room and a mausoleum. Büchel's installation was modelled on a drug clinic from the artist's home country where the very matter of war – the ashes of its victims – replaced the drugs, and the clinic led into a waiting area with an empty space where users would feel the effects of what they had just experienced. Büchel is known for his creation of 'hyper-realistic' environments – fictitious, yet very believable renderings of situations that are constructed so that the exhibition or gallery context is removed. And this particular exhibit was no different. The experience was as bizarre as it was horrific. To move through a single door from the neutral and thus comfortable exhibition space to an entirely convincing clinical environment almost without warning, was very disquieting.

Upon visiting this memorial to a re-imagined Iraq War, it is not clear who Büchel's drugs were intended for. In this respect, the exhibit is evocative of another temporality of warfare – not just in terms of the suffering of civilian victims but also that of war veterans, including those who suffered from 'Gulf War Syndrome' during the First Gulf War. Intruding into Büchel's clinical experience of the consequences of war, gallery visitors had the opportunity to watch the unreality of CNN – synonymous with the 1991 Gulf War – playing continuously on a single TV monitor in the waiting room.

The waiting, drug-infused patients one can imagine here would pose a stark counter-memory to that of the Iraq War being recycled on the screen before them. In this way, Büchel's work highlights a very different temporality, or decay time, of the human body against the superficial effervescence typical of much of the news coverage that is so antithetical to contemporary war's remembrance.

Büchel's installation thus embraces and challenges the media theorist Wolfgang Ernst's conceptualisation of archive space being liberated from its containment. Digitisation has facilitated a switch 'from archival space into archival time' (Ernst 2004, p. 52). However, Büchel has reframed the museum gallery to a place from which memorialisation might be constructed. This new location has nothing to do with official commemoration and instead uses artistic space to counter the seduction of museum culture. The memorialisation of the digital present thus facilitates a removal of the barriers between the past and the present and redraws how we can imagine the commemoration of war. As a result, the traditional vehicle of memorialisation, the archive, retains an analogue location but has moved from the dust-filled backroom of the specialist historian and instead finds itself in the white walls of the gallery space.

Like those artists who make use of digital media to more directly interact with their virtual memorials, national museums have started to make use of media that allow visitors to decide how they will engage with an exhibition. This makes it possible for a visitor to dip in and out of a display, self-directing their experience and what they will take from their visit. For instance, the Crimes Against Humanity (CAH) exhibition, which opened at London's Imperial War Museums (IWM) in 2002, offered an astonishing contrast to the Holocaust Exhibition that had only been installed just two years earlier. Whereas the Holocaust Exhibition told a chronological story of the past, the CAH gallery did not contain a single artefact or museum object. Rather, it provided a narrative of other twentieth- and twenty-first-century genocides entirely through the use of a thirty-minute documentary projection on to a large screen, and six touch screen consoles that provide access to a database. The result was that the CAH exhibition was more easily 'negotiable', giving the visitor the opportunity to choose what to look at rather than be chronologically told what to make of the past. This in turn reflects the way that online communities engage with the past through their social media feeds, rarely taking the time to exhaustively research a subject but instead allowing themselves to be swept along by the speed at which content moves across their timeline. The ephemerality of the CAH gallery was further demonstrated following the IWM London's £40 million refit in 2014 when, after only twelve years, it completely vanished from the museum.[6] So, if the IWM's visualising of the Holocaust gives us a definitive 'recorded memory' – of an event unique, incomparable and appropriate to representation through a relatively fixed narrative – then the CAH exhibit offered an unfinished set of histories and potential memories that have similarities with digital commemoration in online spaces.

The premediation of memorialisation through the work of artists like Delappe and Büchel has also affected how contemporary war is represented within national museums. For example, the Turner Prize-winner Jeremy Deller's exhibit of a rusting wreckage of a car from the bombing of the historic Al-Mutanabbi street book market in Baghdad on 5 March 2007, killing thirty-eight and wounding many more, was displayed at IWM London in 2010. The car's twisted remains were salvaged to show the devastation wreaked on everyday life by asymmetric warfare. What was striking was this exhibit's placing in the museum's main vast atrium of killing machines from the world wars of the twentieth century. Here the twisted, burnt-out wreckage of the car contrasted starkly with the gleaming and pristine machinery of war that so dominated the museum's central space at least until the museum's 2014 refit.

Both the ICA's memorial to the Iraq War and Deller's work offer stark interventions that are indicative of the uncertainties, anxieties and unease brought about by some of the wars fought and mediatised in the twenty-first century. They are also reflective of a group of artists who occupy online as well as more traditional gallery spaces who have effectively premediated the official commemoration of war. As these analogue and digital spaces are brought into discussion with each other, they have the capacity to reframe the memorialisation of war in ways that are not subject to official national debate. As such, these artistic interventions are out-of-sync with the relatively slow commissioning and construction of more familiar commemorative practices.

The Schematisation of Radical War

A schema is a framework or concept that helps us organise and interpret the world around us. These mental models represent shortcuts and standards that the mind forms from past experiences to help us understand and assimilate new experiences. As the digital present has come to take on ever greater memorial force, schema become harder to construct but take on even more significance. For the more connected and proximate wars are to people's experience, the more people seek refuge in the past. These anxieties create the urge to relentlessly turn over the past in an attempt to find glimmers of continuity and stability. The political effect of this is found in the ambition to reconnect people to a 'collective memory' and thus a shared sense of community. This process is intimately framed by an interaction between the radical past and the memorialisation of the digital present and forms something that we call the schematisation of war.

The schematisation of war helps make sense of the velocity and volume of the billions of images of war that have now suddenly become available. The term has a long and influential history and is based on work by Frederic Bartlett (1932) and the neurologist Henry Head (1920), both of whom wrote about the psychology of memory. Bartlett saw the central process of individual remembering as the introduction of the past into the present to produce a 'reactivated' site of consciousness. In this way, 'the past', for Bartlett, is not some kind of fixed object or phenomenon as such, but rather what is crucial in remembering is our 'organisation' of past experiences. So, crucially, memory is seen as dynamic, imaginative, directed and shaped in and from the present.

In these terms, schematisation includes the identification, retrieval and placement of past images, icons and events on to the emerging present, acting as indices against which the dynamic paradigm of remembrance can be understood. Thus, given war's intimate and inextricable relationship to memory, the schematisation of war shapes both how memory is conceived and the ends to which it is put. This is the underpinning of Winter's two memory booms, each following, although in different ways, the scarring enormity of the World Wars. It is also a feature of the 'globalisation of Holocaust discourses' (Huyssen 2003), where this mode of making sense of the world still holds a powerful memorial trajectory over how to make sense of emergent events.

The uses and effectiveness of schema are revealing as a measure of the scale and impact of emergent catastrophes. On 9/11, for example, US news media struggled to impose an immediate and unambiguous template to anchor interpretations and potentially appropriate responses to the terrorist attacks on the mainland United States. For example, Clément Chéroux considers how media coverage of 9/11 was defined by an 'essential *topos*' of the Japanese attack on Pearl Harbor in 1941 both through image comparisons and through iconographic rhetoric (Chéroux 2012, p. 263). The schemata of war also played a part in the appointment of David Blight to the 9/11 memorial advisory group. For Blight was a historian of the American Civil War, a war that had immediate impact in terms of how he sought to make sense of the attacks against the Twin Towers and the Pentagon on 9/11. As Blight explains:

> We immediately look back for some kind of hook, some kind of marker, some kind of place in the past that will help us understand what is happening to us. After all, without that, we're lost, we're lost in time, we're lost with only the present and the future, which is very, very frightening. And hence at that moment … the constant analogies in the immediate aftermath of 9/11, not only to Pearl Harbor, but to Antietam – the bloodiest single day of the Civil War.[7]

Through schematisation, then, wars bring with them a number of nodal events that afford a measure and scale of experience, anchoring the sense of shock, insecurity and loss of a new emergent catastrophe. Consequently, the past and present are constantly entangled. They form a structural part of the dynamic of commemoration. When invoked, these moments offer the chance for a future and the hope of survival; security through historical distance and an awareness that others have survived catastrophic times.

Schematisation also contributes to the battle over the legitimacy of twenty-first-century warfare and of responses to terrorist threats and attacks. Winter argues that developments in media have 'multiplied the images of the damage weapons cause and the suffering of non-combatants in a way never seen before'. Citing Andreas Huyssen in support of his view, Winter (2013, p. 51) goes on to claim this has fed a growing 'mix of scepticism and aversion to war' such that

> We now commonly see the twentieth century in light of its failings, not as a past that triggers nostalgia, but one that haunts us with its demands for legitimising the contemporary polity in light of the multiple suffering of victims of crimes against humanity, of state terror, of racism, ethnic cleansing, organised massacres and the wide-spread violence of postcolonial independence.

However, the volume of images about war does not necessarily equate to a recognition of what suffering is and what it means in relation to the politics of intervention. Nor does it help with a diminution of the glorification of warfare such that war no longer becomes 'an acceptable way of settling differences' (Bell 2008). Instead, by schematising war through the imposition of templates of earlier conflicts, MSM offers comfort and continuity to those who prefer to understand the present through a kind of re-shooting of history. Thus, photojournalists, picture editors and other news workers assert a mainstream schematisation of what warfare looks like. This is not a new phenomenon, but what is of note is this persistence of twentieth-century icons of warfare amid the abundance of imagery produced out of participative war (see Appendix) as it is reproduced online. For in the context of social media, twentieth-century schemata appear to be redundant.

And yet, as Michael Shaw demonstrates, the recurrence of a particular image from the war in Afghanistan demonstrates how twentieth-century schemata continue to frame our perception of war.[8] The image that Shaw draws attention to is by three leading photojournalists, James Nachtwey, Louie Palu and Tyler Hicks, whose photo stories were published within a two-week period in January 2011 in *Time*, *The Toronto Star*, and the *New York Times*, respectively.[9] All employed a very similar image of wounded US marines in the rear of a military 'medevac' helicopter being airlifted out of the Afghan warzone to safety. Shaw's investigations led him to find a number of similar photographs published in MSM in 2010, 2011 and 2012, all of which he believed revealed 'a stunning display of American chauvinism given the intimate framing of the war in such a redundantly heroic narrative, all eyes on our warriors as saviors on high. And then, what does it mean that such high-profile redundancy can occur with hardly a notice?'[10]

What is most interesting in terms of memory, however, is that this type of imagery is already very familiar to some Western, and indeed global audiences. According to Simon Norfolk, these images are not only synchronic to Afghanistan in early 2011 but are also schema drawn from a long mainstream photojournalism trajectory of the medevac image, embedded in earlier US wars. Norfolk argues that this includes David C. Turnley's 1991 World Press Photo of the Year, which is a 're-shoot' of the iconic Larry Burrows' Vietnam photograph that made the cover of *Life* magazine in April 1965. The result according to Norfolk is that '[t]he photographers are photographing the same thing: they're re-photographing a picture that was made 50 years ago. Those pictures from the Vietnam War are 50 years old.'[11] It would seem, then, that MSM offers its consumers a comfort blanket for making sense of war. This makes different wars instantly recognisable as war, providing continuity of Vietnam, Gulf, Iraq and Afghanistan, despite the political and military inconsistencies of these conflicts in terms of their motivations, (il)legitimacies and outcomes.

Another explanation for the persistence of an American imaginary of the Vietnam War is, according to Viet Thanh Nguyen (2017), the fact that the US owns and controls the 'industry of memory': 'How America remembers this war is to some extent how the world remembers it' (2017, p. 108). Thus, a Hollywood-infused persistent MSM schemata, based on a twentieth-century US memory of war, continues to colour how emergent wars are seen, made sense of and legitimised.

In the twenty-first century, however, wars are participative. Anyone with a smartphone can record and upload an image of war to social media. These images rarely conform with an MSM schemata that has its origins in the twentieth century. Consequently, imagery in the twenty-first century challenges and contaminates traditional spaces of remembrance and memory. This juxtaposition jars perspective between individual and collective memory, the perspectives of global MSM and the demands of national politics. The extent to which these individual visions of war might challenge Nguyen's 'industry of memory' is part of the emergent battle between digital memory and history.

Accelerated Memorial Discourses

The radical past and digitising the memorial present come together to form part of the new war ecology. This is a space in which the past is being invoked and denied in response to a new post-trust form of politics. Supercharged through social media, older twentieth-century schemata of war have to compete with online schemata, where national framings of war struggle for attention with wider transnational imperatives. In the mix between old and new, we see how recorded media constrained history to particular forms of representation. At the same time, social media breathes fresh air into older modes of commemoration, reframing how these traditional forms of remembrance are interpreted through an online lens.

In the twentieth century, the capacity to record, store and disseminate history was a function of the analogue technologies that were then available. Depending on the editorial choices of those working in national broadcast media, the net result was that considerably less of the world was being recorded. Historical distance was consequently built into the media itself. For not only were recordings scarce but they also came with machinic, magnetic, film and artefactual decay, degradation and loss. Thus, the scarcity and fragility of the media afforded value to the past, making it worthy of careful excavation, re-imagination and representation. The attitude towards these precious memories emerged out of a scarcity culture that understood and valued the limitations of broadcast media.[12]

It is important to remember, however, that there is a double scarcity to much of the broadcast era's media from the twentieth century compared with the digital modes of engagement today. First, everyday life was not routinely or systematically 'published', recorded or disseminated in an indiscriminate fashion; and, secondly, comparatively little of what was recorded found its way into accessible archives. Indeed, as an examination of the BBC archives will reveal, even television programmes that had been broadcast to audiences of millions were not always kept.[13] Prior to the datafication of the everyday, then, there had been no way to extensively trace time through media. Instead, the default position of the broadcast era was the decay of what had been recorded.

Thus, our immersion in the present flux of the digital contrasts with the decay time of analogue twentieth-century media. In these circumstances, the decay time of these older recordings can only be elongated through digitisation. This prevents media from becoming unreadable as a result of technological obsolescence and in the process contributes to today's ever-expanding digital media ecology. Of course, the vulnerabilities of older media can still be found in digital form as a result of hacking, deletion or accidental loss. However, the counter to media decay remains the digital archive, the apotheosis of which can be found in the cloud.

Where twentieth- and twenty-first-century media clash and compete is predominantly and typically framed by the archivists and museum curators who are intimately involved in deciding what is preserved and digitised and what might be left to further decay. Given the relative glut of analogue material and the limited resources available for digitisation, the possibility of total remembering gives way to selective choices over what is digitised. The result can be an excessive overindulgence in particular aspects of war at the expense of more salient or societally controversial considerations. So, for example, the UK government has denied and then been slow to release the Hanslope archives, an archive of 1 million documents from the Foreign and Commonwealth Office that has the potential to offer fresh insights into British imperialism.[14] At the same time, the First World War has been excessively, if not obsessively, publicly historicised and commemorated. The result, as Simon Jenkins notes in *The Guardian*, leads one to conclude: 'Enough of Remembrance Day ... The composite of the Last Post, "lest we forget" and "Oh! What a Lovely War" is impregnated with enmity, atonement, forgiveness and self-congratulation. It has been reduced to the compulsory "corporate poppy".'[15]

More corrosively, remembrance from this perspective can be seen not merely as a benign feature of our current relationship to the past, but rather as a matter of settling grievances. Consequently, the existence of the Hanslope archive makes it possible for Kenyan survivors of the Mau Mau insurgency in the 1950s to gain compensation for torture administered by British armed forces (Bennett 2012). At the same time, over-remembering also stokes conflict and prevents memories from being buried. Jenkins continues:

> Almost all the conflicts in the world are caused by too much remembering: refreshing religious divisions, tribal feuds, border conflicts, humiliations and expulsions. Why else but for memory does Sunni fight Shia or Hindu fight Muslim? India and Pakistan seem unable to get over memories of Partition. What ancient grievances motivated Myanmar's viciousness against the Rohingya?[16]

The Indian sub-continent aside, accelerating memorial discourses have also had explosive results in Western countries that have yet to come to terms with differing representations of their past. A good example of this can be seen in the public rage over the legitimacy of Confederate statues in the United States. Most of these statues had been erected between 1890 and 1929 and were part of a conscious effort to rewrite the history of the Confederacy so as to justify a system of racial segregation known as the Jim Crow Laws (Domby 2020). Embodying this process were the statues of Confederate leaders and generals like President Jefferson Davis and Stonewall Jackson in Richmond, Virginia. Although many of these were pulled down as part of the Black Lives Matter (BLM) protests in the summer of 2020, for nearly a century a racist past had festered in plain sight but struggled to find any secure presence in mainstream narratives.[17]

BLM may have struggled to gain attention in MSM; however, in online contexts the #BlackLivesMatter hashtag, first used in 2013, drew activists together on social media.[18] This call to action became an online rallying point for mobilised communities not just in America but across the world. Debates that previously were not happening within the political establishment were being forced on to the public agenda through force of direct action, coordinated virtually. Unfortunately, protests and deaths were unable to bring about change. So, for example, the killing of Heather Heyer and the injuring of dozens more after a car was driven into them as they protested against a white supremacist rally in Charlottesville, Virginia, on 12 August 2017 was unable to reframe debates. Instead, President Donald Trump was equivocal in his response, suggesting that blame was shared between the alt-right extremists and Antifa before going on to condemn 'hatred, bigotry and violence on many sides'.[19]

Many historians support the protestors who want to bring down the symbols of the Jim Crow Laws. At the same time, the radicalisation of memory that has been embodied in the protests also poses a challenge for the profession. For as the American Historical Association (AHA) put it, 'what is the role of history and historians in these public conversations?'[20] If historians are to have a place in these national debates, then decisions about memorials 'require not only attention to historical facts, including the circumstances under which monuments were built and spaces named, but also an understanding of what history is and why it matters to public culture'.[21] As James Grossman, the executive director of the AHA, says, the problem for historians is that

> [o]n the one hand, we see an apparent tolerance for historical ignorance, but on the other, there is a renewed national interest in the subtleties of history and memory. Large portions of the American public – including the media – don't know the history of the Confederacy, the Civil War, or how many of the Confederate monuments came to exist. So they are debating whether or not to erase a history that they don't know very well.[22]

That this silence or denial was left to historians to explain reflected the fact that history is being consumed by the politics of memory. This has led historian David Blight to argue that the Civil War is American history's 'sleeping dragon'.[23] This sleeping dragon represented an 'existential Civil War, fought with unspeakable death and suffering for fundamentally different visions of the future'. A conundrum for Blight is that despite the repeated 'shock of events' of 'racial reckonings', the removal of Confederate monuments and the steady stream of parallels drawn with the past, America has never been 'collectively prepared' to face up to its past.[24] Instead, the legacy of a settled but challenged Civil War memory has forced historians into defending the value of their work by countering the fakery and flux of a 'post-trust' (Happer and Hoskins 2022) American president with the certainty and security of an intelligible past. The radicalisation of memory has consequently brought out the dynamic interplay between history and memorialisation.

Thus, this radicalisation of memory – where the past is being put to polarising and exclusionary ends – reflects the wider new war ecology where extreme views are rewarded and outrage is celebrated. This has proved to be a fertile place for state-backed organisations like Russia's Internet Research Agency to manipulate social divisions through social media. This included using Facebook to set up racially divisive events during and after the Charlottesville protests in order to stir up hatred.[25] All of this points to a convergence between a seemingly ungovernable and archive-arresting social media and a radical reworking of memory in such a way that history fades out of view in the relentless recycling of the moment.

Reflecting on the Radical Past

The radical past is made up of three constituent parts that establish a dynamic paradigm of commemoration. In the first instance, the radical past has an analogue dimension that existed prior to digitisation. Typically, this reflects the politics of commemoration as constructed through the nation-state. Secondly, the annual commemoration of key events is interlaced by special anniversaries and less popular memories of the past, all of which find themselves stacked together in new digital contexts. Finally, the radical past exists in a dimension that has been enabled by the digital present. This reconfigures commemoration in ways that are not a function of national politics and reimagines the purpose of commemoration along lines defined by online communities.

These cycles of commemoration do not, however, exist separately from wider discourses about contemporary war. Rather, the digital era ensures that those engaging in constructing and debating memorial discourses cannot help but come into contact with those fighting in and commenting on ongoing war. On the contrary, contemporary wars are both enfolded in and rub up against memorial discourses. At the same time, the rapid digital reproduction and accumulation of data in relation to war itself generates further iterations of all past wars. Thus, the experience of Radical War occurs at the nexus between these phenomena, between those who participate in war's online reproduction and those involved in the accumulation of digital representations of wars past. This is particularly noticeable in online spaces like blogs or social media.

Radical War reflects this blending of old and new modes of memorialisation, where there is a 'tendency to fold past within present' (Lowenthal 2012, p. 2). The urge to find an immediate sense of meaning leads to a relentless churning of war's representation, leading to a constant re-examination and repositioning of the past. This process is being driven by the availability and presence of war-related stuff, which in turn helps to underpin the impulse to commemorate. The result has seen an extension of the roll of museums and artists into online spaces such that commemoration is premediated by online, activist and artistic critics of twenty-first-century war. If we look backwards, we are still drowning in the persistence of the schematic memory that came out of the twentieth century's two memory booms. As we head deeper into the twenty-first century, however, the manipulation of doubt is being accelerated in the perpetual prisms on what is and what is not seen.

Radical War feeds off this fragmentation, where the past is made more slippery as it is invoked, denied and fed by digital infrastructures that promote a new post-trust politics of polarisation, division and exclusion. Governments regularly seek to contort narratives in order to divert and distract (Rid 2020), but the current radicalisation of memory is different from earlier moments or periods where the past and present have folded into or collided with each other. In the twenty-first century, the configuration of the internet and the networking platforms that it makes possible are perpetuating an information ecology that encourages echo-chambers and information prisms. When set in this context, the decay of the traditional archive in favour of something accidental, contingent and subject to information war, guarantees that memory will have an even greater role in framing the past than it has during previous memory booms. This represents the final triumph of memory over history in a culture in which history's stock value of some semblance of facts, and trust in what facts are and what they mean, is put into meltdown through changing digital infrastructures.

5

THE WEAPONISED ARCHIVE

Building an Islamic state that 'is better at taking care of its citizens' demands good record-keeping (Sheikh 2016). As a proto-state, IS understood this and put effort into maintaining its archives. If IS had ignored this aspect of state-making, then its leaders would have been unable to properly manage resources or maintain law and order. Good record-keeping ensured appropriate taxes could be collected fairly, in a timely manner and in a way that stood a chance of gaining the population's allegiance. This was important because taxes made it possible to pay soldiers who in turn could help to expand and defend IS. The larger the state became, the more complicated governance became, and the more things needed to be coordinated in writing.

Maintaining good records and taking a highly structured approach to administration creates a sense in which IS had permanence. Records meant history, and history is where legacies are built. With its borders under threat and its history on the verge of being rewritten by its conquerors, IS's media operators turned to the new war ecology to avoid extinction.[1] They also promote legacy messages that remind participants about the positive world that IS was trying to bring about for the Muslim community. 'Archivist amplifiers' like the Upload Knights (Fursan al-Rafa in Arabic) find ways to slip beheading videos and IS leadership speeches on to YouTube.[2] Here the goal is to keep a sense of nostalgia for what was the possibility of a caliphate[3] to ensure that memories of what once was can be kept alive in the minds of those who have been motivated or who might be motivated to help rebuild IS at some point in the future. In this respect, IS propaganda wasn't just about the immediate demands to motivate its followers, defend IS from counterpropaganda or even to intimidate its enemies (Ingram, Whiteside and Winter 2020). Rather the goal has also been to sustain the idea of an Islamic state even if its geographical location disappeared.

In this respect, IS has consistently demonstrated an appreciation for the interplay between the MSM and social media (Williams 2016). Inside IS, the media strategy focused on maintaining control over the media ecology. Here the goal was to build a digital divide between those people living outside IS and those citizens who lived within it.[4] For citizens of IS, access to enemy propaganda had to be limited. That involved limiting the availability of online media while using broadcast media like the radio or 'media points', which act as open air cinemas, to distribute IS propaganda. For those remote parts of IS, mobile 'media points' were used like agitprop trains.

All of this contrasts with how the rest of the world was exposed to IS. Here IS favoured a participative approach to war such that '[e]veryone that participate[s] in the production and delivery [of propaganda should be regarded as one of the IS's] "media mujahidin"'.[5] IS has used gruesome and highly provocative acts of extreme violence in an effort to guarantee that its social media presence feeds into Western MSM reporting. This has played into IS's media strategy, which focused on portraying Muslims as victims of Western hypocrisy while emphasising how IS was building a better future for its citizens. The result has alienated the Western public even as it has provoked and amplified positive messages among IS's core Muslim audiences.

The explosion of violent imagery and IS-inspired propaganda represented a serious challenge for those trying to limit the effectiveness of IS's media strategy. IS-inspired data traversed multiple information infrastructures, grabbing attention along an arc of unknowable data trajectories, reaching audiences in a whole variety of different ways. This starkly revealed how propaganda moved at a greater pace than Western governments seeking to counter IS propaganda and put social media platforms on the frontlines of the war against IS. At the same time, it also demonstrated that social media companies no longer had control over the content that had been posted to the platforms that they themselves had created. For Andrew Hoskins and William Merrin, this is part of an emergent 'military–social media complex', of the mostly US-owned and developed technologies, platforms and apps, being weaponised across the globe, including against them and their allies (Hoskins and Merrin 2021).

That social media platforms could not control the content of their own sites tells us something about the digital infrastructures upon which contemporary society depends. At the same time, it also helps to explains the interest in machine learning and AI. For these tools are essential for enabling content managers to control or identify inappropriate or criminal content faster than it can be reposted. During the early phases of COVID-19, for example, Google decided to make use of machine learning instead of human moderators for comments on YouTube videos. This prompted a Twitter storm following revelations that critical commentary of Chinese Communist Party (CCP) YouTube videos was automatically being deleted. This prompted Google to state that this was 'an error in our enforcement systems' rather than a matter of company policy or a decision to sweeten relations with the CCP.[6]

In these 'new' circumstances, then, it becomes apparent that there is an ongoing battle between actors trying to control content on the internet and those who want to wrestle control away from the big web platforms. The accelerating cycle between those trying to publish and those trying to censor reveals the weaponisation of the archive. In this digital straight to publication environment, the sanctioned and the illicit, the sanitised and the exposed, the benign and the toxic, all clamour for attention in the live and continuous battlespace of Radical War. This makes a memory of warfare less about its likely trajectory towards a settled, social understanding of the past and more about the digital politics of an unsettled present. This is driven by the connective effects of contemporary information infrastructures, where communications media and archive collapse into the everyday world of the social media news cycle. The consequence is an unlimited capacity for instant playback that does not permit silence or encourage human imagination or natural reflection for thought or forgetting.

Thus, war is caught shifting dimensions 'from the era of recorded memory to that of potential memory' (Bowker 2007, p. 26). In this context, the archive cannot offer a neutral calibration for framing how to interpret the transfer from recorded to potential memory. Instead, the archive becomes a source for reinforcing inferences about the past. The reinforcing inferences have taken on a life of their own, stuck in a permanent loop and unable to achieve the kind of socially accepted understanding of the past that was possible in the age of analogue broadcasting. From this perspective, remembrance in the twentieth-century context can be caricatured as possessing a discernible trajectory of representation and sedimentation that passed through the decline and decay of print and magnetic tape in which it was first captured. However, war fought in today's era of potential memory is locked into a perpetual prism. The multiplicity of competing visions in rapid and continuous circulation of any one event, and the connectivity and contagion from a particular moment, feed an unquantifiable number of informational loops.

But in an era of potential memory, where all recorded data can be used to track and trace individual targets, the archive also offers the possibility of infinite targets. These targets can be identified through an interrogation of what has been recorded, revealing hidden networks depending on who someone has been associated with. In this respect, then, the archive is weaponised, representing a double move in which potential memory reinforces inferences from the past and at the same time forms the stuff that continuously feeds the needs of those looking for a conspiratorial adversary. Thus, the place of the archive in Radical War is not limited to how twenty-first-century society constructs history but is also directly engaged in the process of redefining how we understand and construct the idea of the enemy.

In this chapter, we set out the parameters and the effects of the digital archive in shaping and reshaping the experience, memory and history of warfare, but just as importantly, how this also frames the way adversaries are identified and enemies constructed. We consider the archive as a critical technological and cultural force somewhere in between individual and social remembering and forgetting, and one that alienates memory and history from their traditional trajectories towards becoming accepted social facts. This reveals how wars fought, captured and networked in the digital era – that have emerged through and are now inseparable from the digital infrastructure – inhabit a more unsettled, more clearly contested existence compared with those forged in a pre-digital era where the framing of the enemy appeared to be more sedimented into a collective consciousness. And this in turn points to the way that social media platforms bring analogue, MSM and digital media into more immediate archival relations as they shape how targets are made in the twenty-first century.

Reconfiguring the Digital Archive

The archive has long been seen as the supreme medium, as the external and institutional basis for the remembering and forgetting of societies at different stages of development across history, and as an ultimate storage medium and metaphor of memory. But today the archive is networked, connected, mobile: in short, it is available 24/7 and carried everywhere. The archive can be said to be radical in its anti-archival effects. Rather than containing its objects, its contents, the digital archive expands their connections, their reach, their apparent accessibility. Thus, for example, as Michael Moss and David Thomas consider:

> Brewster Kahle dreamed he could archive the internet, but . . . we will argue that he will wake up one day to find that the internet has archived him. Far from being an object that is archived, we will argue that the internet is itself an archive, but one which does not conform to the rules of archiving as we know them (2018, 118).

Moss and Thomas see the digital archive 'not as a utilitarian warehouse for digital stuff, but as something sublime with extraordinary potential to challenge the way in which knowledge is constructed because, as David Weinberger (2011, p. 61) asserts, it scales indefinitely' (Moss and Thomas 2018, p. 118).

The digital archive does not operate in the same way as traditional, more spatially bound archives. This is partly a function of how connected devices allow their owners to record, store and share information about their everyday lives. Cloud-based photo collections or music streaming create a sense that people are in control of their own archival practices. The digital archive thus seems to make every moment appear endlessly relivable and unforgettable so that today's events no longer have a 'once through' quality of the inevitable passage of chronological time. Rather, they are inextricably part of a politics of 'new memory' (Hoskins 2004, 2018) in which all experience is potentially subject to rewind and to challenge. In this way, the function of the archive is reversed as it is no longer some kind of repository of evidence or verifier of what was, but instead becomes a means to open up the past to a perpetual post-trust state of speculation, denial and counterclaim.

The archive accordingly provides both the political and the infrastructural basis for extending the potential for how societies both remember and forget. First, there is too much of the past that appears to be available to us. This unlimited supply of information is treated too easily as knowledge. As a result, the digital archive upends evidence and expertise through its scale, immediacy and abundance. This has produced an information ecology that facilitates the rapid spread of post-truth or post-trust propaganda, which in turn undermines the perceived value of history and of the work of historians, the effect of which has provoked the AHA to leap to the defence of their profession.[7] Although the AHA's statement responded to the controversy surrounding the sudden visibility of Confederate monuments and the commemoration or otherwise of America's racist past, the underpinning context is the swirl of digital post-trust archival culture in which anything goes.

A central concern for the AHA is the significance that should be afforded to 'evidence' in the making and challenging of historical claims. And yet, to reiterate Moss and Thomas's words, the digital archive in its complexity and scale possesses 'extraordinary potential to challenge the way in which knowledge is constructed' (2018, p. 118).

Secondly, and at the same time, even as the accumulation of uploaded, liked, commented upon, shared or simply ignored materials form an astonishing glut of data, little or no comprehension is paid to its future ownership or finitude. Thus, digital archives are radical in that they mess with the idea that archives are in some way permanent, that there is an equivalence to a physical document or a magnetic tape located in a storeroom. In this respect, the transition to digital archives is not some mere upgrade of what went before but rather a fundamental unsettling of the relationship between the passage of time and the processes of decay that are normally associated with physical media.

Hosted across multiple servers, the digital archive is often assumed to be permanent, immutable beyond the possibility of doubt, with all this data persisting and forming what is routinely treated as a permanently accessible source of information. However, as anyone who has lost their cloud-based photo-archive will attest, there is no basis for any kind of confidence in the persistence of anything held digitally. Data decays, files get deleted, hyperlinks break, cyber-attacks corrupt code, criminals use ransomware to prevent access to material. This adds to the sense of doubt that comes from working in a new war ecology where false information spreads more quickly than truthful content. Most users do not understand the way that their data is encoded and view the inner workings of their digital devices as incomprehensible magical black boxes that should be left to the software developers. This feeds uncertainty and creates seemingly unlimited potential for information war. At the same time, the lack of understanding about how digital images, videos, emails, messages and the possible amalgams of digital content circulate and are reproduced creates even more doubt. In this context, if we see contagion and spread as key criteria for effective informational warfare, of likes and shares, of what Peter Singer describes in *Like War*, we must ask 'where does war end?' (Singer 2018).

And our answer to this question is that war does not end, that in Radical War it is more useful to see the battlefield as always live, being reproduced, recast and reframed in ways that draw connections between history, memory and contemporaneous events. The result is an uneasy convergence of memory and history as they become forged from the same seemingly limitless circulation and accumulation of weaponised content. When will, for example, the gunman's livestreamed video footage of his attack on New Zealand mosques in 2019 disappear offline and/or lose its terrorising power? Will it remain online forever, will it be reframed for a different purpose or will it disappear into obscurity? The answers to these questions, on both counts, are unlikely to be knowable in the foreseeable future. Certainly, when it comes to the fallen image, there does seem to be a diminishment in the sensitivity to, if not the power of, images of suffering in war over time (cf. Sontag 2003; Hoskins 2004). But in this respect the lack of a gap or silence between the moment of an event's publication and its later availability for replay and review messes up our capacity to engage with war's horrors. Marianna Torgovnick sees the difficulties in being confronted with mass violence and death, in shortening periods of time, as part of an emergent 'war complex'. This for Torgovnick includes the work of '[t]he holes in the archive, the ellipses constitutive of cultural memory ... forming patterns that make a certain intuitive sense' (2005, p. 7).

In archival terms, then, there are radical possibilities for the perpetual replaying of wars made and fought and experienced over digital networks, including new opportunities for ellipses, as Torgovnick would frame it, and for erasure. The online prism through which conflict is provoked and seen becomes a relentless evocation of war that plagues and constantly feeds doubt. This sense that digital archives are permanent surely creates an ever-accumulating test for society's capacity for remembrance, commemoration, celebration and mourning, or in David Rieff's terms, its capacity to forget (Rieff 2016). That the digital archive is also subject to decay, nevertheless, messes up the traditional schematisation of war, undermining the notion of a sedimented cultural memory, and this inevitably affects how we conceptualise, justify and come to understand war.

Thus, the digital archive upsets this ordering of cultural and collective memory. For what we are now witnessing is how most communicative acts – be these everyday or extreme – leave digital traces that in effect have become part of the digital archive. Additionally, communicative acts also follow multiple and accumulating trajectories, as messages, emails and comments. And as war is in part constituted through these informational trajectories, the digital archive itself becomes a critical infrastructure of war. But in these circumstances, what are war's limits? What can be remembered, historicised and learned from the perpetual battlefield of social media compared with those traditional battlefields that were delimited in time and space? Another way to configure these questions is to ask what kind of war is being produced from and hidden in this new archival force? It is precisely amid the perception of being overwhelmed, of war's radical potential for being replayed and remade over digital archives, that a new form of war is being fought – one that involves fighting over the archive itself. And it is to this form of war that we now turn our attention.

The Archive at War

The potentially infinite digital reach, spread and duration of war, scattered and connected in new ways in memory and history, offers a range of opportunities for the fabricating, leaking, sharing, hacking and stealing of information on an unprecedented scale. Most digital archives do not sit in isolation but are dynamically constructed in relation to other connected data systems. Consequently, as Arthur Kroker observes, the digital archive is 'not only recording and responding to online queries and new links but effectively adapting its future behaviour to that which is trending on the net' (2014, p. 89). And it is precisely this organic and distributed process of adaptation, spread out across billions of internet users, that challenges the where and the when of contemporary warfare. For war is in constant evolution with the archive that hosts it. The digital archive appears to be continually unfolding, adapting and accumulating. In this respect, there is a sense in which it captures everything. In this way, the digital archive seems profoundly transparent. And yet, paradoxically, in another way the scale of the internet also illuminates our inability to encompass and control the stuff of the archive. Geof Bowker (2016) captures this tension:

> Where we are left is inhabiting an uneasy zone between a proximate future, always just around the corner – about five years away, when we will have all the data – and a set of archival practices in the present that perpetuate certain kinds of invisibility: things we cannot or will not see.[8]

The idea that archival practices perpetuate a kind of invisibility of things that can and cannot be seen seems counter to the notion of the accessibility of the digital archive and the transparency posed by participative war (see Appendix). This includes the apparent democratisation of the uploading of a panoply of voices and videos from the frontline. Yet it is precisely the overwhelming immediacy and accumulating volume of digital media that make it impossible to contemplate online spaces as having any sort of orderly organisation in what archivists would otherwise understand as an indexed and annotated archive.

In this respect, given the nature of participative war, the archive is not something that is created, stored, ordered and retrieved in the traditional sense of the relationship between the present and the past, but is rather constituted 'on-the-fly'. This happens through the uploading of content or offering commentary on what is already on the web. Here social media is constantly feeding from and responding to a multitude of users, offering a mediatised form of collective memory. This kind of open-source material is supplemented by the kinds of continuous surveillance afforded by drones in what Chamayou calls 'a revolution in sighting' (2015, p. 38). The combination of these source materials produces an avalanche of recorded data that might be used for the purposes of tagging targets. The sheer volume of data nonetheless constitutes the supreme challenge in our model of Radical War in the struggle over data, attention and control. AI has the potential to be a key weapon in helping to reveal meaning out of this data. At the same time, however, attention is premediated through the application of AI to data analysis and prediction, potentially leading to the prioritisation of targets that mirror the prejudices of the software designers who themselves must constantly rewrite their applications to identify hidden adversaries.

The corollary of the weaponised archive, then, is that Radical War forms out of a perpetual beta or prototype war in which war is continuously made and remade on the battlefield, in digital infrastructures and reconstituted through the identification of new patterns out of established archives. The speed at which this must take place, if enemies are to be countered by new doctrines and weapons, produces an impetus towards prioritising experimentation and adaptation in real time, while in battle, rather than at the lab or on the training ground.[9] This has parallels with those working on software applications in the cyber and digital domains, where the idea of fielding prototype systems as some sort of beta war has a long history.

Thus, for example, the work undertaken on armed drones was instrumental in ushering in the most recent wave of beta war. Having its origins in the late 1990s, the US Predator programme was developed by a group called 'Big Safari'. They were something like a tech start-up and had long specialised in adapting traditional Air Force aircraft for secret and time-sensitive operations.[10] What was interesting about Big Safari was their willingness to put into action work before it was fully tested. As Arthur Holland Michel explains, '[t]he team referred to this as "the 80 percent solution" (because sometimes the last 20 percent of a job takes the longest)'.[11] This, according to Brian Raduenz, the head of Big Safari's Predator group at the time, was like releasing the beta version of a piece of software.[12]

Similarly, and as we explained in the last chapter, Radical War can be conceived as war in a perpetual state of beta testing. Surveillance through the weaponised archive forges a constant premediation of the threat: the continuous pre-emption of the enemy being seen as everywhere, helping to frame evidence, shape intelligence and construct approaches to targeting. As a result, as Chamayou puts it: 'The archives of lives constitute the basis for claims that, by noting regularities and anticipating recurrences, it is possible both to predict the future and to change the course of it by taking pre-emptive action' (2015, p. 45).

A key technology to this end is Wide-Area Motion Imagery, or WAMI, which is a significant advancement over the narrow fields of view of traditional video cameras that needed to be pointed at the suspected target. Inspired by the film *Enemy of the State*, WAMI is capable of filming entire cities from above, watching and recording wide areas to identify and track hundreds of people and vehicles moving in real time even over more than 100 sq. km.[13] Multiple video feeds can be viewed by different operators from within the camera's vast field of view, including sending automated alerts if something or someone of interest moves into a particular area. At the same time, the technology allows the entire area view to continue to be recorded, even while a segment is being zoomed in on. In this way, Bowker's 'proximate future, always just around the corner – about five years away, when we will have all the data' (as cited above) may have arrived. WAMI is nothing less than a total memory machine, an all-seeing eye, what in Michel's terms might best be described as an all-seeing archive.

This then enables a new kind of pure war – to borrow from Virilio – in which time itself is arrested and the archive is weaponised, over and over. John Marion, for instance, outlines this new archival extended present of war:

> [I]n addition to its real-time monitoring, a WAMI system acts like a time machine. Most system configurations record, tag and archive all imagery collected during their time in the air (hours, days or weeks). Users can then 'rewind the clock' and conduct forensic analyses of significant events over an extended period, all while monitoring ongoing activities in real time. This technique can uncover significant relationships between people, vehicles and locations that might otherwise have gone unnoticed.[14]

Mirroring the possibilities opened up by WAMI, the lifelogging device named 'Narrative' promises to take a similar approach to your memory. Here the goal is to link your photograph collection to key memories, allowing you to organise, search, retrieve and relive your experiences. 'Relive your life like you remember it', its marketing video pronounced: 'The camera and the app work together to give you pictures of every single moment of your life, complete with information on when you took it and where you were. This means that you can revisit any moment of your past.'[15] Some computer scientists have developed this idea further and are advocating a Human Digital Memory (HDM) through the combination of an array of data and content types. The Digmem system, for example, gathers various data from smart, connected appliances (Dobbins et al. 2013) in such a way as to allow users to search through their past memories. This has led one of its proponents to suggest that '[i]n the future you could simply ask, "When have I been happy?" And the system would return all the information associated with that emotion.'[16]

The cultural logic of HDM, and the related belief that securing a kind of total memory is possible and desirable as a way of managing the past, mirrors the military's belief that the gathering and aggregating of more data will deliver a future memory of adversaries that will enable predictive and precision targeting. The datafication of the everyday feeds systems of surveillance that create, tag and archive the digital individual. In turn, the excessive and continuous production of data creates an archive of past activity that is a constantly available resource that can be repeatedly mined for new insights about the network of people who form part of the digital grid. The digital archive thus becomes the epicentre of Radical War in that it enfolds individuals into a myriad of potentially unlimited data manipulations in which they are knowable and targetable. This produces a shift in how we think about warfare. For the abundance of data makes it possible to identify targets before they realise that they are being targeted. Under these conditions, warfare hardly represents a duel between fighters in combat. Rather, according to Chamayou, warfare is being rewritten such that it more closely represents a hunt defined by pursuit (2015, p. 52). This shift in the perceptual field has reframed how enemies are labelled and conceptualised, and it is to how the vision of an enemy has changed that we turn next.

Beheading the Fallen Enemy

In the wake of the attacks of 9/11, George W. Bush launched the United States on a war against terrorist groups with global reach.[17] Later renamed the Global War on Terror, the American attempt to defeat terrorists wherever they might be found ushered in eighteen years of what the *New York Times* subsequently described as the forever or endless war.[18] It has become known as the forever war because al-Qaeda has proven difficult to pin down and defeat. Indeed, President Bush warned as much when he told Americans they should 'not expect one battle, but a lengthy campaign' in which there would be little opportunity for 'the decisive liberation of territory'.[19] The frustrating aspects of the GWOT had their antecedents in the twentieth century. The First Gulf War had spelt out that if an enemy of America fought in the open then they would make themselves easy targets for the US military. To survive, it would be essential to avoid attraction. This meant finding ways to hide, sometimes in plain sight.

The decision to defeat terrorists wherever they may be found has had a number of second- and third-order effects. One of these has been to catalyse the way Western powers conceptualise the enemy around which war can cohere. Thus, as the academic Debjani Ganguly puts it, the 'post-cold war change in the idea of the enemy from measurable and identifiable to immeasurable and irrepressibly fractured has irrevocably shifted the ground of traditional warfare, both rhetorically and strategically' (Ganguly 2016). Moreover, as the anthropologist Allen Feldman argues, this has had implications in that '[t]he departure of a reliable enemy seriously threatens war as a system of political-discursive commensuration and capitalisation that a calculable and prognosticated enemy secures and anchors' (2009, p. 1705). This idea is neatly summed up by the way that war was declared against a concept – as in the GWOT – rather than a traditional enemy.

More corrosively, however, within the space of twenty-five years, the clarity found in the idea of an iconic and evil leader, someone that can be identified as *the* enemy, against which militaries and political and public support can be roused, has unravelled. Saddam Hussein, for example, may have been one of the most demonised leaders (at least by the US) in the modern age, but his continued appearance in the news media called into question the finality of the outcome of the 1991 Gulf War. As he led United Nations weapons inspectors on a merry dance during the 1990s and the early 2000s, the Iraqi leader's existence increasingly cast a long shadow over the failings of US foreign policy. And it was the White House's hawkish memory of their failure to remove Saddam from office in 1991 that drove them to obsessively demonise him. The subsequent 2003 invasion invoked a memory of the First Gulf War that had been kept conscious by a news media reliant on a twentieth-century televisual archive of mostly stock images of the Iraqi leader (Hoskins 2004).

The point about Saddam being defined as *the* enemy had two results. First, invoking Saddam meant obfuscating evils that might be found elsewhere in the world. Second, by shaping American strategists' imagination, it fuelled hugely unrealistic expectations that equated the downfall of Saddam with the 'liberation of Iraq' and peace in the Middle East. Brian Walden, the former British politician turned political commentator, argued prophetically in 1999 about the consequences of these representations of the Iraqi leader, saying:

> Look what our demonisation of him has led to. It's an extraordinary story. That so much suffering, so much human and material cost could have been inflicted upon the world by somebody who's little more than a bandit chief is amazing. Here we stand: with all our technology, with our computers and our smart bombs and our global economy. And there he stands: a creature almost of another world leading a shambles of a country which has never been able to solve any of its fundamental problems. And yet he manages to come out on top. Why? Because we refused to accept the complexities of the real world and understand them, and decided to impose upon events the stirring simplicities of a Hollywood action film. We've done it before, and we shall probably do it again.[20]

Indeed. And ultimately it was the bloody aftermath of Saddam Hussein's demise, including his messy execution, that showed how embodying the evil enemy in a single identity effectively smooths over many of the entangled geopolitical, cultural, religious and historical complexities that the continuation of his presidency – no matter how abhorrent – kept from exploding into civil war.

Ultimately, the logic of demonisation in this period of Western interventionism is bound up in the logic of regime change. For example, in February 2011 protests began against the rule of Muammar Gaddafi, leading to an armed uprising and the Libyan

Civil War. In March, UN Security Council Resolution 1973 demanded a ceasefire and authorised a multi-nation force to protect civilians through military intervention and the imposition of a no-fly zone over Libya. As Hugh Roberts argues in an essay entitled 'Who said Gaddafi had to go?', attempts to secure a ceasefire through negotiation were 'deliberately rejected' given that regime change was the undeclared but hardly secret agenda of particularly the US, France and the UK.[21] The prospect of the sight of people from Gaddafi's regime in face-to-face talks with members of the rebellion was unacceptable to the Western powers as it would have undermined their characterisation of him as someone who should not be engaged with.[22] In these circumstances, Gaddafi's end became the overriding objective for those Western nations who, Roberts argues, used NATO as their proxy to pursue him. Roberts continues that '[s]ince the issue was defined from the outset as protecting civilians from Gaddafi's murderous onslaught "on his own people", it followed that effective protection required the elimination of the threat'. In practice, this involved the targeting of Gaddafi himself for as long as he was in power, which was subsequently revised to 'for as long as he is in Libya' before finally becoming 'for as long as he is alive'.[23]

But a media-accelerated obsession with the fall of a single target as a focus for or legitimisation of a military campaign often leads to an anti-climax, and as with regime change, a vacuum and/or a shift in objectives. For instance, the former Iraqi president emerging dishevelled from a hole in the ground in December 2003 ran across global news networks, and the chaotic torture and death of Colonel Gaddafi, some of which was captured on cell phone footage in October 2011, marked the beginnings of New Wars rather than heralding their end. At least the death of Osama bin Laden in May 2011 delivered some kind of prescribed closure to the post-9/11 US war against the Taliban, although perhaps that was owing to it being publicly unseen (and un-see-able) so the event of his death could match the mythical status of his life. Yet as Jonathan Mahler writes in the *New York Times*, despite the clarity of the US coup in bin Laden's death being muddied with claims as to how exactly he lived and who knew of his whereabouts and the nature of his burial at sea, '[s]ymbolically, it brought a badly wanted moment of moral clarity, of unambiguous American valor, to a murky war defined by ethical compromise and even at times by collective shame'.[24] Unfortunately, this was only a blip in the trend in the fall of the reliable enemy, for as Rony Brauman puts it, '[i]n Afghanistan and Iraq in particular, we have long stopped knowing which war we are waging'.[25]

Identifying the Archival Enemy

As the demonisation of the iconic enemy that dominated much of the twentieth century's reporting of war began to fade in the glare of twenty-first-century media, another enemy was needed to explain why success had not been possible. Not only would this legitimise the actions of Western armed forces; it would also justify the money spent on them. But if the decapitation strategy – whether that was in relation to Saddam, Gaddafi or bin Laden – hadn't worked, then how would planners find the visceral targets that might sustain war in the twenty-first century? What was needed was a new enemy that could be permanently conjured at the beck and call of the military–social media complex, 'generating threats without limits' (Masco 2014). The advantage of participatory surveillance was that it made it possible to create a permanent enemy. All you had to do was mine the limitless archive of data about individuals, their movements and their relationships to find the targets of the future. Total memory, as we have conceived it, offered the possibility to construct infinite targets in what might be described as a kind of total enemy.

Constructing the total enemy has meant bending the boundaries between real-time monitoring and potential memory through the application of intelligence analysis tools such as Palantir Gotham. Founded in 2004 by Peter Thiel, Alex Karp and alumni from PayPal, Palantir took its name from the omniscient crystal balls in J. R. R. Tolkien's *Lord of the Rings*.[26] Playing an instrumental role in the targeting activities of US Joint Special Forces Command, Palantir software enables intelligence analysts to crunch and fuse data quickly, thus making possible the Find, Fix, Finish, Exploit and Analyse missions we discussed in Chapter 3. It combs through 'disparate data sources – financial documents, airline reservations, cellphone records, social media postings – and searches for connections that human analysts might miss. It then presents the linkages in colorful, easy-to-interpret graphics that look like spider webs.'[27] This kind of data association effectively represents an analytical tool for instrumentalising the digital archive for targeting purposes.

While the utility of these techniques has been over-promised, the ambition behind them is clear. And in this respect, the game changer in the generation of infinite targets emerges out of the digital generation of identity. This is not about what you say or who you are, but is rather identification based upon what you do, where you go and who you communicate with. This is finding, paradoxically, the unknown individual. Targets of drone strikes do not have to be known by name but are instead identified by what is called pattern-of-life activity, in other words metadata movements, networks of contacts and mobile phone use. Attacks on this basis are known as 'signature strikes'. So, whereas the idea of a person's signature is usually defined through the act of signing one's name to something, or the name of a person written with their own hand, here it is based upon the signature of metadata and their pattern of life.

For example, the US National Security Agency's SKYNET programme of mass surveillance of Pakistan's mobile phone network, collecting metadata (and only metadata) from 55 million users, used a machine learning algorithm to rate the likelihood of a given person being a terrorist. So, if your pattern of life metadata corresponded to that of known terrorists then you become a target. What the incoming present and the accumulating past of Radical War have in common, then, is that they are both newly searchable, drawing on archives of metadata that are not obviously accessible to those who have created these digital footprints. With WAMI, there is an emergent digital simultaneity of surveillance in real time and in recent time. In the process, it makes use of digital power to order memory in premediated ways based on a recent past that is held in archival suspense all designed for the purposes of locating an enemy.

This trend has been matched if not superseded by the collection of biometric data during the war in Afghanistan. This sought to fix identity against biological data, essentialising individuals to their place in the database (Jacobsen 2021, p. 141). By collecting detailed biometric data 'on 80% of Afghanistan's 25 million people', the American programme was seen as an important way to locate and classify people (Jacobsen, 2021, p. 242). This would enable American forces to identify social networks that could in turn be targeted by security forces. Following the American withdrawal from Afghanistan, this same dataset fell into the hands of the Taliban, who could then use it for the purposes of building their own kill/capture lists.

Framed this way, the difference between the iconic enemy and the archival enemy is precisely that the latter could be anyone. If visibility and knowability were criteria for the symbolic enemy of broadcast-era war, then it is truly Donald Rumsfeld's infamous 'unknown unknowns', of invisibility, that have become principal characteristics of the infinite enemy in Radical War. Thus, a crisis of representation has emerged in which the demonisation of the iconic enemy through the MSM no longer works in the face of a growing concern for conspirators seeking to destroy society from within. These conspirators must be defeated if victory in war is to be delivered. In this respect, the process of rendering transparent the myriad invisible connections through data mining and visualisation generates potential enemies of convenience that can be targeted pre-emptively and even anonymously. Radical War can thus be fought against no one and against everyone.

The idea that the state is trying to identify the enemy within has not only created new enemies but as a by-product it has also fed the trolls from QAnon and the Boogaloo movement who inhabit Facebook, Reddit and 4chan. In the case of QAnon, despite West Point's Combating Terrorism Center describing the group as a 'bizarre assemblage of far-right conspiracy theories', President Trump found himself unwittingly endorsing their members.[28] This in turn has further fed conspiracy theories associated with the 'deep state' and the claims that Trump is fighting an international cabal of metropolitan elites.

But it is the demonisation of the headless hydra that is Antifa that really tells us something of the place of conspiracy theory in the new war ecology of the twenty-first century. For Antifa has played a significant role in mobilising former President Trump and his supporters to bring about a radical political realignment in the United States that has the potential to reshape America's entire strategic outlook towards embracing civilisational war in Steve Bannon's 'The Fourth Turning'.[29] As protests against the murder of George Floyd broke out in cities across America, the threat to use the Army against protestors led to an officers' revolt in defence of the American constitution and against the president.[30] Trump's use of Antifa as the ultimate enemy within had reached its highwater mark but in the process revealed the importance of conspiracy theory not as 'a symptom of resignation, as critical modernists would have it' but as a driver 'of cultural transformation in the West' (Aupers 2012, p. 31). In this respect, Americans are watching the GWOT collapsing in on itself, from externalising the threat following 9/11 to Trump withdrawing from overseas entanglements and then having to construct a hidden enemy to keep the terror going.

All of this was happening at a time when 22.1 million jobs had been shed between January and April 2020 and the unemployment rate had climbed to 14.8 per cent as result of COVID-19 lockdown measures.[31] Inevitably this recalibrated anxiety levels among those communities left most vulnerable by the financial crash in 2008. In this context, the COVID-19 pandemic has provided a perfect opportunity to take beta war into the heart of society so as to further entangle the individual in the network and the archive of infinite targets. Thus Naomi Klein argues that

> [i]t has taken some time to gel, but something resembling a coherent Pandemic Shock Doctrine is beginning to emerge. Call it the 'Screen New Deal.' Far more high-tech than anything we have seen during previous disasters, the future that is being rushed into being as the bodies still pile up treats our past weeks of physical isolation not as a painful necessity to save lives, but as a living laboratory for a permanent – and highly profitable – no-touch future.[32]

Klein sees the pandemic as offering the optimum opportunity for government and big tech to collaborate over 'a future in which our every move, our every word, our every relationship is trackable, traceable and data-mineable'.[33]

The COVID-19 crisis provides the perfect storm for Radical War. After years of conceding access to our personal data in exchange for increasingly frictionless digital living and working, populations are at their most vulnerable, having become subjects in the mass experimentation of participatory surveillance. The enemy produced out of this weaponised archive will not be the demonised leader, branded and simplified as good versus evil for largely anonymous audiences by the MSM. Instead, the enemy becomes whatever and whoever emerges out of the digital archive, as part of a mundane process of identifying and targeting individuals.

As we argued in Chapter 1, it is the habituated use of the smartphone – and the promise of friction-free convenience and personalisation – that makes us less conscious of, or less willing to fully engage with, the threat posed by the datafication of everything that is mundane, including where we go and whom we meet. It is the smartphone that offers up entire populations as potential targets, making it a structural feature of Radical War's battlefields, battlefields that we willingly construct through our own participation. Yet while it emerges through the often seemingly dormant but instantly available digital archive, it is persistent real-time video surveillance and near-instant recall of data imagery that premediates how memory and history in the twenty-first century operates. These premediated interpretations do not happen by accident but emerge through the eyes of the electronics and software engineers who construct these systems of systems.

On the one hand, the demise of the iconic enemy is in part a function of the instability of some post-Cold War states and their descent into complex civil wars. At the same time, the symbolism of evil embodied for years in convenient enemies through mostly obliging Western mainstream news media has also been undone by the emergence of non-traditional media in the new war ecology. Thus, the fall of this enemy is also a symptom of what we have set out as a crisis of representation. But the wars in countries such as Iraq and Libya that followed the removal of their dictators and governments is not just a symbol of a failure of military interventions and twentieth-century Western conceptions of civil–military relations. It also reflects and has been enabled by fundamental changes in the representations of war in the digital archive and as it is framed in total memory. This socio-technical assemblage is not just a consequence of the relative weaknesses of state power in the international order but more accurately is indicative of the success of Western 'civilisational warfare' (Brauman 2019) over the information infrastructures of those who now depend on them.

PART 3

CONTROL

Diagram 5: Control

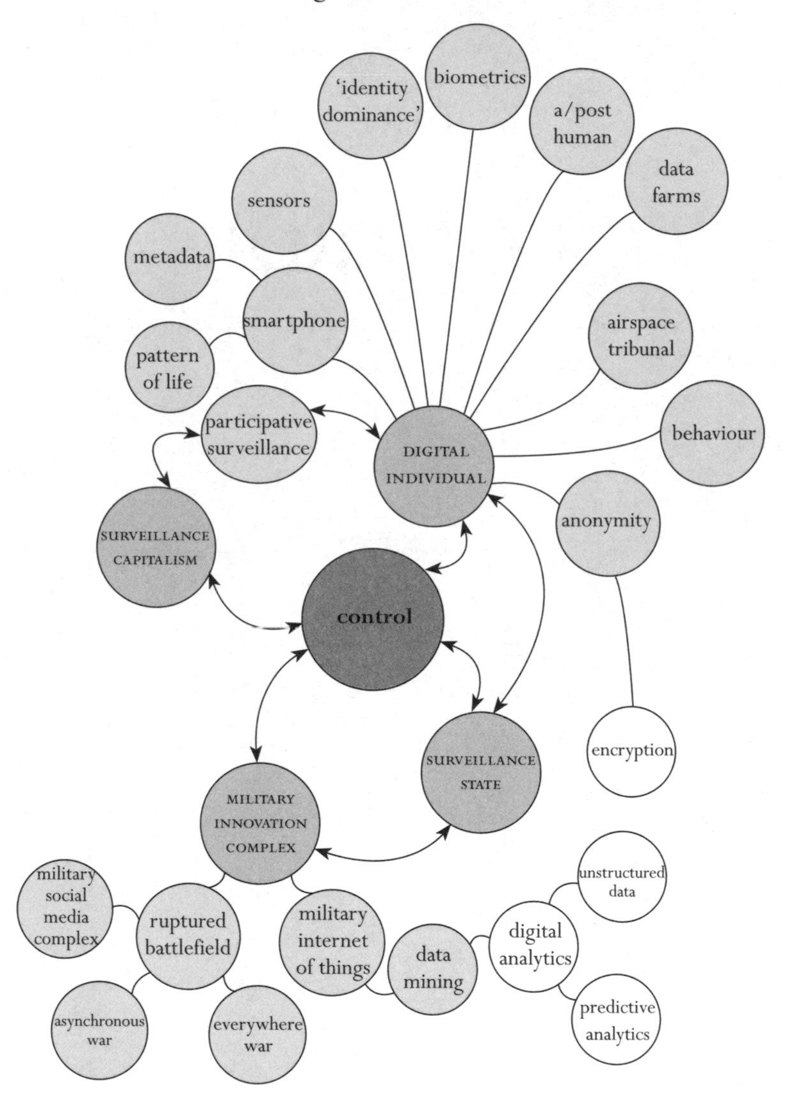

6

TECHNOLOGIES OF CONTROL

COVID-19 contact tracing apps offer a useful lens through which to view contemporary technologies of control. The challenge facing the developers in France and Britain was that Apple and Google could determine how these apps would work and whether personal data could be shared with government health agencies. Elected government struggled to persuade the Silicon Valley firms to relax the privacy settings on their operating systems, effectively leaving Apple and Google with the capacity to 'understand and intervene in the world, while truthfully saying they never saw anybody's personal data'.[1] Big tech could use their infrastructural power to dictate to sovereign governments how personal data would be used.

In the battle for control over the relationship between data and attention, much depends on the uneven distribution of the informational infrastructures of the new war ecology. Government may be sovereign, but big tech has the means to drive social change. Big tech created the transnational infrastructures that made possible the multiple data trajectories that have imploded our understanding of war and deeply mediatise how we experience the world. However, in their evolution, the information systems that enable twenty-first-century war have been unevenly distributed across the globe. This unevenness recasts digital divides in all sorts of ways, creating perfect opportunities for military exploitation and upending older, well-established categories of war by refashioning the way we produce knowledge about battle.

In this new war ecology, the war for control over these infrastructures stretches from the battlefield to the technology platforms that enable and co-construct contemporary experience. It is not easy to see how this 'code/space' manifests itself until the technologies of control are themselves revealed for what they are (Bridle 2019). Thus, it will not be until augmented reality game Pokémon GO becomes a phenomenon in Moscow that we will more easily be able to see that the Kremlin is surrounded by a GPS distorting field that prevents players from acquiring Pokémon and foreign agents from taking GPS coordinates for the purposes of tracking and targeting. Only through these everyday processes of rupturing can this reality be exposed, revealing how the digital world is reshaping social relations at home even as it exposes new dimensions in questions related to war.

In the twenty-first century, information infrastructures are remodelling the military just as Uber rewrote how we order a taxi and Airbnb helped us to find a holiday rent. Put simply, Radical War is not managed by the military alone. It cannot be compartmentalised into neat doctrinal, physical, information and cognitive domains (Alberts et al. 2001) because the new war ecology is always on, and always participative: its infrastructures constitute the very pathway through which we frame our understanding of war. Inevitably, this has disorientating effects on the military bureaucracies who continue to organise their activities along twentieth-century dimensions associated with what is inside the state and what is external to it (Walker 1992). The effect is to emphasise how technologists put in place the organisation and systems necessary to help the armed forces operate in these new ways. Unsurprisingly, it also has the effect of putting technologists in the driving seat for shaping the way that the armed forces think about war.[2] What is emerging is not the warfare of early European history, where the prospective state is but one actor among many others trying to gain hegemonic sovereignty through military power (Tilly 1985). Rather, what we are watching is a battle for control between those who design and build information infrastructures in the US in Silicon Valley, in China in Zhongguancun and in Russia at Skolkovo (Dear 2019; Kania 2019; Lewis 2019) and the state bureaucracies struggling to shape and direct the technologists of the fourth dimension towards the battlefields of the twenty-first century.

The effects of this process can be seen in all sorts of places, including the fight for people with the necessary skills to design, build and maintain these contemporary information infrastructures.[3] It is reflected in the exorbitant cost of housing in North California's Bay Area and the corresponding homelessness in downtown San Francisco.[4] But it can also be seen in the militarisation of these infrastructures across society. This includes the transfer of technologies from the public into the military domain, the application of archiving tools for crowdsourcing military history and the re-tooling of veterans as software engineers. Thus, veterans of the GWOT, for example, now code javascript like they are at bootcamp.[5] In the US military, mIRC, a free and open-source type of Internet Relay Chat, represents the 'weapon of choice' for flattening the military's strict hierarchy and enabling communications between commands, security levels and alliance partners.[6] And in the creation of new citizen military historians, Operation War Diary uses Zooniverse, an online web platform that helps to crowdsource the study of First World War archives, massifying user research experience for the purposes of academic-public engagement.[7]

Nowhere is this process of militarisation more obvious than at a DoD-sponsored hackathon. Although non-military hackathons date back to the early 1980s, the US DoD was concerned that American forces had lost their technological edge and instigated a series of initiatives in 2016 intended to bring talented software engineers, material scientists, PhD students, academics, management consultants, venture capitalists and warfighters into the same orbit. Unable to develop AI itself, the DoD is reliant on companies like Google. But with Google employees objecting to working on projects for the US military, it is clear that the US government has much to do to persuade the San Francisco tech giants to direct their energy towards US strategic goals.[8] This in turn has prompted serial tech entrepreneur Peter Thiel, the founder of Palantir, to comment that what was 'Good for Google' was 'Bad for America'.[9] As unlikely and faddish as these two-day events might sound, then, these hackathons are socially engineered to convince America's knowledge workers of the importance of developing American information infrastructures that solve challenges defined by the US military. As a result, civilian 'tech' nerds from elite colleges get to meet frontline troops and discuss how they might design technology to address a battlefield problem, advance the 'mission' and save lives.

The hackathon may not conjure the same sense of awe that seeing a twentieth-century aircraft carrier shoot and catch F-18 Hornets from the flight deck may provoke, but it is important not to underestimate the immense cultural power that these events represent. Framed as a response to outdated modes of acquisition, where bureaucracies take at least five years to go from idea to implementation, the hackathon epitomises the digital silver bullet to outmoded, overly hierarchical and unresponsive civil–military structures.[10] In this respect, the hackathon is not just an innovation event but also symbolises an attempt to completely revolutionise the way that the US DoD does its business. The goal is to refashion the ethos of US defence acquisition in directions that were pioneered by the networked entrepreneurs that evolved out of the revolution in micro-computing of the 1980s. Rooted in a culture that has its origins in the counter-cultural hacking and hippy community, these entrepreneurs sought to escape conventional and broken governance structures and refashion society through networked personal computers along more direct and participative lines (Turner 2008). The hackathon thus acknowledges that the US DoD does not own processes of digitalisation but must instead respond to how technological change has rewired the social experiences of those engaging in government. As such, the hackathon represents an attempt by policy-makers and Silicon Valley influencers to bring the digital revolution into the last of the analogue, hierarchical and bureaucratic bastions of the twentieth century.

Given the ambition that lies behind the hackathon, creating workable technologies that might solve battlefield problems is in many respects secondary to the underlying agenda that frames the emergence of these events. Building a network of highly capable people outside the DoD is an explicit object of the conference organisers. For the real goal is to reflect digital experience into the bureaucracy while shaping the thinking and capturing the imagination of future technologists in what might best described as a war for talent.[11] Put simply, as far as the US DoD is concerned, there are not enough talented people working on the interface between the military and the digital challenges that civilian tech entrepreneurs of the early twenty-first century have created for contemporary warfighters. If technologists are to be attracted into defence-related challenges, the question facing hackathon organisers is how to direct talented people to conclude that their future lies in working with the military–innovation complex rather than on directing themselves towards alternative societal or commercial projects in the fourth dimension.

The question in this respect is not about how the media manage the agenda, process, channels or instruments of communication (de Franco 2012). The participants of war are, in large respect, one and the same. This is then a question of how to win the war in the cluttered mental space that Virilio describes as the grey ecology (Virilio 2009), to capture what is now an 'economy of attention' (Goldhaber 1997; Wu 2016). How then are experiences engineered to grab attention 'when attention is scarce and information abundant' (Broadbent and Lobet-Maris 2014) and digital infrastructures are largely beyond the control of any single actor? And how are the information infrastructures – whether that starts with engineers erecting telecoms masts or AI designers trying to maintain control over the web platforms that now occupy billions of people's time[12] – being used to maintain control over the digital divides that sustain them (Ball 2020)?

Manipulating Digital Divides

Gaining attention in a data-abundant environment where 96.7 per cent of the world's population can access a mobile-cellular network, 93.1 per cent can access a 3G network and 84.7 per cent can access a 4G network is one thing.[13] Exploiting the digital divides that typify uneven distributions of information infrastructure is another thing entirely. At one end of the digital divide, we find an ongoing fight for control over the design and development of the meta-infrastructures of the new war ecology. Here the big scare stories fluctuate between models of digital society that are bounded by surveillance capitalism and the surveillance state (Zuboff 2019), framing, for example, the way Toronto's citizens describe social innovation as 'Smart Cities' even as equivalent technologies are being used to oppress the Uighur Muslim population in Xinjiang.[14]

But generating information advantage is not just structured by state ideology; it is also a function of infrastructure investments that go well beyond the control of the state. In this respect, the big nine tech firms of the world – Google, Amazon, Apple, IBM, Microsoft and Facebook in the US and Baidu, Alibaba and Tencent in China – are central, dominating investment in AI, machine learning, cloud-based computing and predicting human behaviours from rich contextual data sources (Webb 2019). Here surveillance capitalism and the surveillance state manifest themselves as technologies that observe and manipulate behaviours, produce fake faces, spoof voices and deepfakes, where software algorithms calculate how an image or video of someone's face can be manipulated into doing or saying something that they never in fact did.[15] These technologies are about recreating digital divides between people experiencing life in an abundant information environment.

At the other end of the spectrum, in parts of the world that have been poorly served by the internet and the technologies that enable it, digital divides are imposed more directly through the hardware itself: by reducing the cost of smart technology but pre-configuring it with a dis-information architecture that reconfigures what can be seen.[16] On the one hand, this is about convenience, connectivity and choice. At the same time, new communication channels premediate attention, which with editorial care can be used for political effect. The Philippines, for example, has been swamped by cheap smartphones.[17] This in turn has revolutionised political campaigning, enabling Rodrigo Duterte to use Facebook as a central platform for his election to the presidency. As part of a campaign that highlighted a tough war on drugs message, Duterte's PR strategists directed digital influencers, community-level fake account operators and grassroot intermediaries to shape public debate.[18] Ultimately unleashing a wave of vigilantism in the police and a breakdown of due legal process, Duterte's strategy culminated in the presidential sanctioning of murder that not only targeted drug traffickers but also drug users more generally (Pomerantsev 2019).

But these digital divides are especially noticeable when you go directly on to the battlefield itself. For it is in the crucible of violence that we can start to pick apart how some constituencies retain a voice while others self-censor or remain silent. Take for example the Syrian Civil War. When it comes to exploring the structure and effect of digitalisation, more than a decade of war in Syria has been catastrophic for Syrians, but the war has proven to be very useful for those wanting to understand and prepare for future war. Thus, we have witnessed the United States perfecting the targeting tools and techniques it had pioneered in the first decade of the twenty-first century. As a result of these efforts, drone warfare, Special Forces operations and kill/capture missions have reached a pinnacle of refinement. At the same time, IS has demonstrated the importance of controlling narratives while manipulating web-messaging for recruiting, branding and state-building purposes (Winter 2019). Consequently, we've seen a mash-up of technological and tactical innovations, where old and new systems have worked separately and alongside each other, combining into and refashioning warfare with off-the-shelf technologies to create weapons that are as much cyberpunk as they are sophisticated (Hashim 2018; Cronin 2020).

In the midst of the battlefield, civilians have had to adapt their behaviours to reflect changing frontlines. Even as access to internet provision through wired infrastructure controlled by the government has remained limited through electricity outages and bomb damage, a decentralised telecommunications infrastructure has developed. Made possible by mobile phone infrastructures that leverage WiMax or wi-fi links from Turkish cities that provide subscription services to local residents or satellite communications to internet cafés, the UN reported that the monthly cost for 40 gigabytes of 3G or 4G data was around US$11 in June 2017.[19] Of the twenty ISPs in Syria, three are government-owned. Independent satellite ISPs have been banned by the Assad government, and cyber-café owners within government-controlled areas require a licence from the Ministry of Interior and are expected to monitor internet users. Lacking the capacity to control a centralised telecoms network, the government only has the partial capability to limit access to web content, censor news, delete data and in extremis switch off access to the internet. As a result, the government must supplement its limited control of the web with harsh criminal penalties, detentions, surveillance, intimidation and technical attack either from the security services or the Syrian Electronic Army, a group of pro-government hackers who have targeted opposition organisations.

The possibility of civilians facing severe repercussions is not limited to connecting to the wrong network inside territory controlled by Assad. The information environments within IS have been just as precarious, and as battlefronts move, internet connectivity frames the behaviours of those charting the violence. For those capturing evidence of chemical weapon use, for example, care has to be taken to avoid uploading material that names victims or makes allegations. If you sign on to the wrong network or that network is taken over by an enemy, naming people may produce unintended negative effects for all those involved. Equally, as the NGO Syrian Archive notes, even when data is uploaded vigilance is needed to ensure that content is not removed. Sometimes content removal and account deletion are accidental. At the same time, manipulating the digital record clearly presents an opportunity to propagandise. For content platforms like YouTube, this puts search content moderation algorithms on to the digital frontlines, which in turn has led NGOs into a race against time to preserve important evidence so that future war crimes might be properly prosecuted.[20]

The instability of the archive, in terms of how platforms like YouTube decide on what content can be posted and retained and because adversaries now recognise that they must try to block or curtail the availability of uploaded material, points towards the multiple challenges associated with maintaining control over the grey ecology. As we have seen, the digital archive purports to be a continuous twenty-four hours a day, seven days a week means for shaping narrative. This puts narrative manipulation on a 24/7 cycle, bringing the war for control directly on to data centres based in locations that are far removed from the battlefield. The extent of the commotion produced by this constant turmoil has been weaponised as prototype or 'forever war'. The upshot is that social media platforms have lost control over their own platform, leading one commentator to observe that 'these crises are developing faster than its minders can address them'.[21] Consequently, social media companies have to predict the technological requirements necessary to keep capturing these experiences, thus premediating the demands of those anticipating future battlefield events. In this respect, just like weapon designers, companies like Facebook must put themselves at the forefront of research and development in relation to AI if they are to retain any control over the information environment that is otherwise subject to appropriation by various political, economic and criminal actors.

The result is a cycle of remediation and premediation where the past as expressed online is disrupted, transformed and diffused in a constant and ever-tightening feedback loop. Although this process had its origins in Web 1.0 and the wired communication networks of the twentieth century, an accelerated cycle of innovation and change has been provoked by the ambition of taking greater control over users' digital experiences. The ensuing collapse of binary categories, between participant and observer, that otherwise helped us make sense of war in the grey ecology has further accelerated the cycle of change in an ongoing and never-ending battle to retain control over content. This in turn has become an end in itself, sucking the big nine tech companies into a vortex of technology development. Here it is hoped that, for example, AI, machine learning and facial recognition will reassert the capacity to identify real-time data patterns and thus regain control over the information tsunami that the tech entrepreneurs of Silicon Valley have enabled.

One apparent way out of this cycle of remediation is to target the information infrastructures that enable data dissemination. At the most exquisite end of this spectrum lie cyber-attacks that either bring down computer networks, corrupt or delete databases or switch off power grids and other essential networks that sustain the functioning of modern societies.[22] As Andy Greenberg observes, the most obvious example involved Ukraine, which he argues constituted a 'lab test' for establishing the effectiveness of this type of activity. At the more direct end of the targeting spectrum lies physical infrastructure such as mobile phone masts and other parts of the wired telecom infrastructure (Berman, Felter and Shapiro 2018) that control and enable data transfer and communication. In the latter case, by physically shutting down the communications grid insurgents and terrorists have been unable to coordinate activity, trigger IEDs using their cell phones, or record and upload the explosions and subsequent ambushes that drive online propaganda. Some academics have drawn the data together and argue that there are strong correlations in targeting physical infrastructure and the dampening effect this has on political violence (Berman, Felter and Shapiro 2018). By contrast, others argue that cyber-attacks in wartime have done little to change battlefield events (Kostyuk and Zhukov 2017). Whatever the reality, physical infrastructures are targeted either with cyber or kinetic attack because the military believe that the distribution of communication networks influences and sustains political narratives that justify war. Thus, by cutting off people's capacity to use these networks, the military can shape systems of competitive control that disincentivise some behaviours while incentivising others (Kilcullen 2013).

And this points to the fact that the armed forces no longer talk about military effectiveness without also discussing audience effects. Just like an online commercial marketing campaign, the goal of a military information operations team is to try to win audiences and shape interpretations of events, with a view either to mask a military action or engineer the public's response to one. In this respect, a military action also has performative force, where the destruction of a bridge is not only a way to disrupt a supply network but can also convey a political message or shape public attitudes. The challenge, however, is to identify a sort of conversion rate, where the relationship between clicks and views equates to a change in behaviours in target audience participation.

In many respects, it is the military's desire to define a correlative effect between a predefined message and a change in audience behaviour that has driven the emergence of an influence industry. The contemporary military approach to influence operations emerged out counterinsurgency campaigns in the Middle East. In this respect, and in contrast to kinetic operations, influence operations were designed to win the 'hearts and minds' of those population groups that needed to be secured from insurgents (Ford 2019). Among American forces, the doctrinal aspects of these activities were designed for the wars in Iraq and Afghanistan, where steps had to be taken to counter the effectiveness of insurgent propaganda (Rid 2007; Briant 2019). Here the challenge came from Iraqi and Taliban forces, who did not suffer from the same hierarchical military structure as the US Army and were therefore more agile at making propaganda by uploading video of successful IED attacks and their follow-up ambushes than coalition forces (Hashim 2018). Having grasped the centrality of winning the information campaign, however, Western armed forces eventually got much better at framing the information environment as they sought to influence the undecided in the hope that they might choose not to side with the insurgency.

The military's use of influence operations to win the strategic communications battle during the wars in Iraq and Afghanistan provided the necessary oxygen for the emergence of a much wider influence industry driven by data analytics. Having started operations and then refined their methods in countries outside the West, this new influence industry has gone on to try to shape public opinion in Europe and North America. With contracts to the UK's MoD, the US DoD and NATO, the most noticeable organisation in this respect was Strategic Communication Laboratories (SCL) and its subsidiary company Cambridge Analytica.[23] Initially, SCL undertook behavioural analysis and overseas influence campaigns aimed at shaping elections or referenda in as many as thirty countries before its activities came to world attention following Britain's referendum on membership of the European Union in 2016.[24]

By the time Cambridge Analytica was being employed by Vote Leave, the official campaign team for the UK to leave the European Union, SCL had perfected a number of capabilities to scrape Facebook user profiles and illicitly harvest other data sources so that they could develop micro-targeted political advertising (Briant 2019). This in turn had been based on earlier research into Facebook profiles aimed at 'tracking the digital footprints of personality', which offered insights into how to tailor messages for individuals.[25] Given the quantity of data that had been harvested, the teams working for SCL employed '[t]hree machine learning pipelines ... to process both text and images. The software could be used to read photographs of people on websites, match them to their Facebook profiles, and then target advertising at these individual profiles.'[26] The combination of academic research and the application of AI for identifying key audiences led SCL to make some bold claims about the relationship between the messages they conveyed and the success of their campaigns. Thus, for instance, while working for the American Conservative advocacy organisation SCL would 'claim that the 1.5 million advertising impressions they generated through their campaign led to a 30% uplift in voter turnout, against the predicted turnout, for the targeted groups'.[27]

Since 2016, the influence industry has grown significantly, with Brittany Kaiser, the Cambridge Analytica whistle-blower, observing that there are now hundreds of companies using similar techniques to SCL working on influence campaigns.[28] Indeed, Buzzfeed reported that as many as twenty-seven online disinformation campaigns with connections to PR or marketing firms had been exposed as fake, with one promising to 'use every tool and take every advantage available in order to change reality according to our client's wishes'.[29] But this is not just limited to political campaigning. On the contrary, one reason this has happened is because businesses have been persuaded that data analytics will help them successfully target potential customers and grow their market share. Consequently, we have seen a growth in sponsored posts by people branded as 'influencers' on social media platforms like Instagram. The size of the influencer market in 2018 was valued at $137 million. This was expected to grow to $162 million in 2020.[30] With the global digital advertising spend topping $378 billion in 2020,[31] of which most goes to Google and Facebook, it seems that a lot of marketers believe that e-commerce driven data-analytics combined with targeted online marketing can produce serious financial returns for those businesses willing to invest.[32]

Whether the likes of Google or Facebook have the tools to hack your brain and predict how you will respond to an online message is, however, open to question. For example, online experiments have shown that targeted personalised ads are typically aimed at an audience that is already very likely to buy a product. That is to say, advertising has limited effects at reaching audiences beyond those who have a pre-existing preference. And if this is actually the case, then much of the money spent on online influencing fails to reach those people a business might want to target if they are to grow their market share. As one commentator observes, though, the fact is that both marketers and those who purchase online digital influence 'believe that their marketing works, even if it doesn't'.[33] If this is true, then the intoxicating idea that digital infrastructures can be used to predict human behaviours does not hold much water.

What is true for the marketing industry is also true for the armed forces more broadly. Indeed, there has been a long-standing interest among American and Russian armed forces – an interest that stretches right back to the Soviet Union's efforts to invert nuclear deterrence and game theory through a process of reflexive control – in shaping perception and directing human behaviour (Chotikul 1986; Thomas 2004). Compared to the Cold War, what's different now is the way that web platforms provide immediate feedback loops on how influence activities are being received by audiences. This implies that organisations like SCL or Cambridge Analytica have the capacity to more effectively gauge response rates; and this is catnip for the armed forces, for the military want this kind of predictive analysis because without it they cannot hope to control interpretations of war in the twenty-first century. Like businesspeople wanting their marketing budget to produce a predictable return on investment, the military are predisposed to wanting a conversion rate on messaging. For if they cannot control the messaging, then the fear is that they have lost control of people's perception of the battlefield. And in this respect, data analytics and digital infrastructures breathe new life into the military's ambition to predict the relationship between message and change in behaviour, even though it is not clear whether these influence operations are likely to reach audiences beyond those already predisposed to the messages being conveyed.

All of these countervailing impulses remind us that the way data is created and captured reflects the deep mediatisation of contemporary war. For the fact is that the armed forces are reliant on data mining and analysis tools that are only made possible by technologists working on platforms produced in Silicon Valley. Recognising government's dependency on the digital infrastructures it provides, Silicon Valley seeks to gain further leverage over bureaucratic data by offering cloud storage and other data management services.[34] All of this private sector activity does of course run in counter-cultural ways to that of the hierarchical and compartmentalised military bureaucracies producing the data in the first place. At the same time, it also opens up the possibility that something as mundane and boring as records management might be made sexier through the addition of machine learning and AI, facilitating analytical processes that sift through vast data stores quickly and draw connections between material that otherwise might be impossible. What is also true, however, is that there will be no going back once government bureaucracies hand over their data repositories and leave themselves dependent on cloud technologies.

All of this poses problems for those powers like China and Russia who are on the wrong side of the digital divide with the United States even as they seek to rebalance world politics away from American influencing activities.[35] As the Snowden files reveal, these wide-ranging influence activities include trying to persuade technology companies to look on US intelligence activities more favourably, even to the point of redesigning US-made servers and routers.[36] In 2013, for example, the US National Security Agency spent $250m a year on a programme that aimed to 'covertly influence' the product designs of technology companies. By March of that year, the agency had already created the infrastructures necessary to gather 97 billion pieces of data on just six countries in just one month.[37] Not only does this reveal the significance of US government investments in cyber-security and digital espionage; it also implies that American information infrastructures are highly advanced – despite their protestations to the contrary.

When American information infrastructures are compared to those of China and Russia, it becomes apparent that all three countries take different approaches to surveillance and engaging their IT communities. Whereas the Chinese government takes a highly militarised approach to its IT projects (Kania 2019) even as it implements draconian surveillance,[38] the Russian government's aspirations for AI, machine learning and robotics far outstrip its capacity to take advantage of venture capital investment or attract or even educate appropriately qualified engineers (Dear 2019). It is important to avoid the orientalist mistake and claim that open societies are somehow better at innovation than authoritarian states, for it is clear that China and Russia are intending to prove Western naysayers wrong in this respect. Indeed, some commentators observe that Russia and China may merge their digital activities at some point in the future and develop formal 'elevating bilateral ties' that outstrip US capabilities (Dear 2019).

This reveals that innovation in the suite of AI technologies that the military intend to use for the purposes of accelerated warfare is also a problem of creating appropriate cultural, economic and social conditions for recruiting highly skilled technologists and entrepreneurs. In these circumstances, the Russians, Chinese and Americans are competing in a technological cold war that is not so much about industrial espionage and cyber-war but rather constitutes a war for talent in which states hope they can lure businesses into working with them. Here the world's big nine technology companies have the whip hand, controlling access to technology. This contrasts with non-state actors who are looking for immediate advantages by applying open-source innovation and generic design plans disseminated on the internet to generate more immediate battlefield results (FitzGerald and Parziale 2017; Cronin 2020).[39] For this battle for talent reveals the uncertainties that frame the lack of progress in the state's war for controlling the digital environment as much as it demonstrates the eagerness of the military to harness technological innovation for use on the battlefield.

Nonetheless, this is not stopping Russia and especially China from taking more direct sovereign control over infrastructures that they can build for themselves and use to form rival influence channels (Segal 2018). For example, in November 2019 Russia brought into force a law creating a sovereign internet, which is designed to 'bring the entire network infrastructure under political control' and allow the government to cut off the flow of digital information.[40] President Putin has also rubber-stamped a bill designed to ban the sale of smartphones without Russian apps pre-loaded on to them, a move that has been interpreted as a direct attack on the commercial viability of US tech giants like Apple who are trying to operate in Russia.[41] Russia has since followed up on its efforts to create a sovereign internet by disconnecting itself from the global internet during tests in June and July 2021.[42]

Similarly, China has long been able to manage the flow of web traffic with cyber-censors manning the Great Firewall of China preventing access to sites and censoring what people discuss on Chinese social media platforms like Weibo (Griffiths 2019). Although this surveillance state has been enabled by companies in Silicon Valley, China's sovereign web has created an environment of self-censorship that does not depend on terrorising populations but leads citizens to believe that government is 'competent and public spirited' (Guriev and Treisman 2019).

But while it would be easy to claim that these 'Informational Autocrats' (Guriev and Treisman 2019) are restricted to non-Western states, it is also true to say that similar information practices are multiplying in a number of European states. Viktor Orbán, for instance, has gone out of his way to control the information channels that are open to the Hungarian public, enabling oligarchs close to the prime minister to close or destroy independent news media and throttle the availability of critical reporting. Thus, Hungary's oldest daily newspaper, *Népszabadság*, ceased operating in October 2016 after its reporters were locked out of their email and the paper's digital archives were erased.[43] And in their moves to open media companies in Britain, the oligarchs behind this move have since made their intentions to restructure the media channels of other European countries very clear.

Orbán aside, in many respects what distinguishes authoritarians in Europe from those in other parts of the world can be determined by their regional rather than global levels of ambition. In what is being described as China's Belt and Road initiative, for example, the Chinese government has embarked on a massive investment in road, rail, telecoms and port infrastructures designed to refashion logistics, regulatory standards and finance between the two ends of Eurasia. Phase 1 of the project is due for completion in 2021, but the timeline for the Belt and Road initiative is not scheduled to end before 2049 (Maçães 2018). While it is unclear whether the programme of investments and building will produce the returns that President Xi Jinping has promised, the effects of this expansion of east–west logistics infrastructure are being felt in many of the countries that have been earmarked for inclusion in the project. For the ambitions behind the Belt and Road initiative are not simply about facilitating movement and communications but are rather aimed at reshaping influence and patronage networks across countries and continents. In this respect, the goal is entirely strategic in its outlook, namely to 'build an expanded "factory floor" along the full economic corridor and across national borders' such that entire supply chains will be linked to the Belt and Road and thus to China (Maçães 2018). And if countries cannot afford to take part in this initiative, then China can offer concessional loans that assure Chinese access.[44]

These activities are producing significant geopolitical effects in terms of maritime navigation and control over shipping routes. As climate change makes navigation through the Arctic region an increasing possibility, China has shown an interest in infrastructure along the Northern Passage, which has led to speculation that the Chinese would commission their first ever nuclear-powered icebreaking ship.[45] At the same time, by leveraging its scientific and financial strengths, China has started to use its influence to shape the policy choices of the Arctic Council. Chinese investments in Greenland and Iceland have been considerable, producing initiatives that have widened the debate over Arctic governance to countries beyond the members of the council.[46]

Similarly, acting as a turning point at the corner of the Indo-Pacific, this puts places like Australia on the frontlines. As a distant but safe democratic haven, countries like the United States and the newly invigorated 'Global Britain' can use Australia to project naval power into the Malacca Strait and along the Chinese Belt and Road. By reimagining the Far East as an Asia-Pacific maritime-centric space, the hope is to duck accusations of neo-imperialism even as the goal is to contain China. As the United States seeks to maintain freedom of navigation in the South China Sea, China seeks to infiltrate spies into Australian politics in an effort to undermine the West's grip on Australia's foreign policy.[47]

Given the willingness of the Chinese government to use infrastructure investments as a way to access information so as to gain patronage and reshape existing social networks, it is understandable that US officials have been so resistant to China's mobile telecoms giant Huawei setting up 5G wireless base stations across the world. This is primarily because China is collecting data in far from transparent ways. For example, Chinese contractors were involved in building the African Union's (AU) headquarters in Addis Ababa. As part of this deal, they were also contracted to build the AU's computer network, which according to *Le Monde* also contained a backdoor to allow data transfers to China's security services.[48]

Although espionage against the AU was subsequently denied by the Chinese government, letting Huawei build the world's 5G network would offer China the potential to gather unprecedented levels of data. With the potential to unlock $12.3 trillion of revenue, 5G technology is in hot demand.[49] For not only does it represent a key enabler for the emergence of the IOT; it is also essential for the introduction of smart homes and cities and will drive change in business, healthcare, connected vehicles and manufacturing even as it stimulates further rounds of innovation through deeper datafication.[50] The level of access to people's lives would not only mean Huawei had backdoor access to spy on foreign powers but even more worryingly would help the Chinese carry out pattern of life analysis that could be used to influence the information ecology for entire population groups. In this context, then, it is understandable why American officials are resistant to Chinese firms establishing market dominance in 5G networks, going so far as to propose nationalising the infrastructure and leasing it back to telecoms companies.[51] However, lacking the resources of private industry, American policy-makers cannot hope to develop their own networks independently of private businesses. Consequently, they can only hope to shape information infrastructures if they are to ensure that their interests are taken care of and the security of their networks maintained.

Overall, then, while there is much talk of AI, machine learning and autonomous vehicles, micro-targeted information operations and digital footprint mapping, it remains to be seen how specific systems will manifest in the military's everyday table of equipment and organisation. Even as information autocrats and the influence industry have sought to show the military how to control the information environment, the military, industry and futurologists are still scratching around, evaluating all sorts of platforms, technologies and systems in the hope that at least one of them might produce war-winning outcomes. The one technology that is being driven forward, a 5G data network, is important for enabling military and civilian tech, but this is a contested infrastructure, the dominance of which may well frame the way future geopolitics unfolds.

This rivalry over technology platforms is not, however, particularly new. Armed forces have regularly found themselves having to decide whether to back a premature technology in the hope that it will bring beneficial military outcomes (Knox and Murray 2001). What is different this time is the scale, complexity and sophistication of the informational infrastructures under consideration. In these circumstances, nation-states do not have the capacity to keep up with the rate of change in the private sector and must instead try to harness industry for military effect. But given the speed and multidirectional nature of digital technological change, it is apparent that there is no obvious response to how to win the battle over the technologies of control. More fundamentally, and considering the benefits that technologists in Silicon Valley reap from the swirling changes and the churn in the information infrastructures of the twenty-first century, perhaps it is unlikely that any definitive response to controlling the new war ecology will ever emerge.

All of this plays into the hands of Silicon Valley tech firms who use the idea of a technology cold war as a way to defend themselves from regulators in Washington, DC. Here the argument is that if big tech is not given free reign, then the Chinese government will be able to take advantage (Wu 2020). In these circumstances, it is irrelevant that companies like Facebook fuelled controversy through their association with Cambridge Analytica during the 2016 US presidential election. Nor should it matter that the trillion-dollar businesses have avoided their competition by buying it up, just as Facebook bought Instagram and WhatsApp and Google bought YouTube (Levy 2020). Any attempt by politicians seeking to regulate big tech or apply anti-trust legislation would be a sure-fire way to undermine private investment in AI and give the Chinese an advantage in its effort to influence geopolitics.

The Digital Dysphoria of Radical War

The possibility of population-level disinformation campaigns is double-edged. At what point in this hall of mirrors does it become possible to distinguish reality from disinformation? This is not just a question for the intelligence services, or for civilians participating in a keyboard war located in distant parts of the world, as it also affects the military themselves. For disinformation on a population-level scale, where participants have already stewed in an information environment maintained by the meta-manipulators of the fourth dimension, will have effects on the way the military recruit, understand their trade and organise themselves for war. Indeed, mirroring the uncertainties experienced in the digital world more broadly, Rosa Brooks observes that there is a very real sense of disorientation among the armed forces over how best to construe their future role (Brooks 2016). This is partly connected to wider and longer-term trends in relation to the importance of the state in world affairs, the use of private military contractors on the frontlines (McFate 2019) and the hollowing out of the state in relation to researching and developing frontline technology (Avant 2005; Singer 2011). But it also reflects a growing sense of distrust in experts and professionals (Eyal 2019) combined with the amplification of fake news and the prospect that soldiers will soon have to fight alongside or be replaced by autonomous vehicles and robots.

This swirling sense of dysphoria has been neatly captured by Charlie Brooker in his 2016 episode of the Netflix series *Black Mirror* called 'Men Against Fire'.[52] In this episode, we see future soldiers fighting to protect displaced villagers from an insurgency by human-shaped but facially demonic feral mutants known as Roaches. Consciously drawing parallels with S. L. A. Marshall's post-Second

World War book by the same title – a book that observed that only 15–20 per cent of US infantrymen fired their weapons in combat – Brooker introduces us into a world where the military has developed the technology to improve soldier performance (Marshall 1947). Through a series of bio-tech implants called 'the Mass', soldiers are able to access a suite of technologies that enhance their warfighting capabilities, giving them more effective situational awareness and marksmanship while managing emotion and rewarding the killing of Roaches through dreams and dopamine hits. Only when the Roaches shine a modified laser device into the eyes of the soldiers do the effects of the Mass on soldier perception start to be revealed. Far from being feral mutants, it becomes apparent that the Mass has mediated soldier perception, corrupting it in such a way as to ensure that certain population groups are represented as demons and open to targeting while others are left alone. While this has the effect of sharpening military effectiveness, it also prevents soldiers from seeing through the narratives that their commanders have presented to them.

When it comes to digitalisation and the combat soldier, however, it does not take much to see how the new war ecology is being reshaped around a vision of battlefield singularity, that point where analogue and digital worlds fuse into one visual, haptic and mental register. Indeed, many of the technologies that Brooker brings to life on the screen are already being worked on. Thus, as Dr Thomas Reardon, the CEO of CTRL-Labs, makes clear, wearables, implants and neural interfaces represent the future of bio-tech and the management of cognitive load.[53] Alongside neural interfaces, the Platypus Institute is working on enhancing human performance by training the neural pathways of people to reproduce the brain states of experts. Branded Human 2.0 by the Platypus Institute, this has already produced the means for transferring expert neural signatures in marksmanship to soldiers learning how to improve their shooting skills.[54] Further down the line, Dr Amy Kruse, the chief scientific officer at Platypus, observes that these technologies have the potential to digitally connect units and units with machines through human–machine teaming, enabling remote devices to be controlled neurologically. And if these technologies are not enough, soldiers and marines are already being given the opportunity to road test mixed or immersive visual augmentation, allowing them to access battlefield data directly to sharpen their experience of combat.[55] Lastly, neuroscientists at the University of Toronto are working on how the brain recalls memories, offering the possibility that memory manipulation will be available as part of a suite of treatments suitable for a number of mental disorders.[56] Far from representing a fantasy, it seems Brooker's science fiction is mirroring reality rather than offering military futurologists a vision of future war.

If this episode of *Black Mirror* is anything to go by, it is clear that technologists will be reproducing digital divides through soldier augmentation (Coker 2012). This will relieve soldiers of the cognitive burden of having to process vast quantities of data by providing some sort of computer implant. At one and the same time, this will give neurologically enhanced soldiers combat advantages over those who do not have these systems as well as locking soldiers into a veil of perception that cannot be escaped without severely rupturing an individual's sense of reality. Put simply, singularity will be deeply mediated with the values, perspectives and ambitions of the technologists that have created it, framing what aspects of war are seen, made relevant and what might be considered to constitute knowledge about the world. This has very deep implications that take us back to the heart of the battlefield where the technologies of control will mirror the interests of those in Silicon Valley, Zhongguancun and Skolkovo.

Indeed, as a metaphor for how class in future societies works, one way of reading this episode is to see the Roaches as members of the working classes, rendered dispossessed and terrified, hunted down by a technocratic elite group who are happily willing to demonise and kill them. Looked at this way, Brooker's dystopian vision is not as far-fetched as it sounds. For in many respects the global technological elites are in effect finding ways to dispossess the working class through the application of information infrastructures that disenfranchise and hold back those in society who cannot engage with the digital utopia that emerges out of Silicon Valley.

Getting the armed forces to move from where they are currently to where Brooker imagines them in ten years' time will, nevertheless, involve significant changes to the military's work practices. On the one hand, these technologies are presented as offering the potential to integrate a suite of systems that enhance soldier performance. On the other, however, they represent a very direct way to rewrite the culture and values of the armed forces. In practice, the adoption of bio-tech and the MIOT (see Appendix) will fundamentally rewire how armed forces go about doing work on the battlefields of the future. Given the way that these systems either minimise the importance of soldiers or completely cut them out of the decision-making loop, these technologies also represent a significant downgrading of the importance of the soldier in delivering military effects on to the battlefield, implying significantly changed hierarchical relations between soldier and officer and technologist. As such, by changing the relationship between the soldier and their work, the new war ecology represents a direct threat to the way that the military currently frame their professional expertise; and by extension this implies a corresponding increase in the standing of those people managing this battlefield singularity as opposed to physically experiencing it.

Not only does this have a disorientating effect on current social relations within the armed forces; it also affects how the services frame their sense of professionalism. A central feature of professional status lies in having autonomous control over the boundaries of the work that a profession does. This involves the production of certain kinds of knowledge that others must learn if they are to join a particular professional group (Larson 1979). Reliant on information infrastructures that are well beyond the core competence of the armed forces, the military must now harmonise their professional skills with those of other actors who have increasing importance on the battlefield. This has started with private contractors who have service-level agreements for maintaining military kit, but like the Russian contractor Wagner and the American company Academi, private organisations also now engage in direct military action either for the state or its proxies. As Tony King observes, this has resulted in a form of professionalism in the armed forces in the twenty-first century that retreated to exquisitely technical and uniquely military questions of doctrine and tactics, techniques and procedures (King

2013), where fighting is the central function of the armed forces. But as Brooks observes, the number of people in uniform who actually do this kind of activity is remarkably small given the amount of training that is dedicated to ensuring the armed forces can fight adversaries. Consequently, most soldiers 'hadn't spent much time doing what they thought of as the essence of soldiering' (Brooks 2016). Instead of fighting, the military find themselves managing information and logistics structures. As a result, there is a disparity between what is trained for and what the majority actually do.

All of this poses a whole range of challenges for how civil–military relations are conceptualised. The new war ecology thus represents something of an attack on existing martial work practices and the knowledge processes by which the military maintain autonomous control over the boundaries that mark it out as a profession. This reflects a process that Richard Grusin describes as 'radical mediation' (Grusin 2015). For Grusin, what makes radical mediation properly radical is not just that experience is mediated but that in the digital age a theory of knowledge and a theory of being are now one and the same. Producing knowledge about war is not just a product of fighting but also the product of a collective experience of battle. And in the new war ecology, knowledge production about war no longer privileges the perspectives of what Yuval Harari calls the military 'flesh witnesses' who have seen combat at the frontline but now involves all those who might otherwise be described as being part of everywhere war (Harari 2008).

Thus, the digital trajectories of the ruptured battlefield cannot be considered an epiphenomenon of twenty-first-century war. More accurately, they have folded back on and recast our relationship to how war is known and made sense of both at the frontlines and within the grey ecologies of those who are nowhere near the kinetic engagement. These feedback loops make war in the fourth dimension truly radical. They do not just collapse categories that are otherwise associated with everywhere war, categories such as geography, distance and time but also force us to critically reimagine foundational seventeenth-century categories that help us make sense of the world. These categories include elementary Hobbesian and Cartesian dualisms that frame our understanding of war and peace, body and mind. Consequently, '[i]n place of Hobbes' strict separation of war and peace, there is a creeping militarisation of politics. And in place of Descartes' strict separation of mind and body, there is instead an image of a human being possessed of instinct, emotion and calculation, all fused together' (Davies 2018). This has produced a nervousness in which individuals and governments live 'in a state of constant and heightened alertness, relying increasingly on feeling rather than fact' (Davies 2018, p. 87).

The Hobbesian solution to alleviating these anxieties was to centralise the management of violence to a sovereign government that could control disorder within the state's geographic boundaries by enforcing the law while retaining the right to engage in war overseas. But the state itself still needed buttressing with expertise that would only come with the emergence of professions like surveyors, engineers, scientists, doctors, lawyers and the military. At home, these professions created a body of knowledge that would build confidence in the contractual arrangements between individuals, businesses and governments by reducing the suspicion that one side or the other might default without being held to account (Davies 2018). Abroad, these professions might help generate the means and materiel for war and the knowledge that might secure peace.

Even more significantly than this, however, these professions also created an epistemological framework that made it possible for society to come to agreement about what constituted a fact. As such, the professions were instrumental in producing the knowledge necessary for sustaining the notion of an objective, impersonal and apolitical body of knowledge and at the same time maintaining the framework necessary to sustain these facts. The result was that these experts constituted the gatekeepers of facts, carefully protecting their professional status and position in relation to generating knowledge about the world.

Since the 2000s in particular, however, we have seen a growing distrust in expertise and the professionals who are largely responsible for maintaining a sense of impersonal, apolitical objectivity. While antecedents for this can be traced back a century or more (Eyal 2019), the belief in the technical management of government has been fundamentally eroded by bankers during the crisis of capitalism in 2008 and politicians who decry that the 'people have had enough of experts'.[57] The traditional point of experts and professions has been to reduce the fear and uncertainty that comes from a breakdown in trust that promises will be honoured. These bonds of trust have been replaced by platform capitalists bringing buyers and sellers together, enabling them to manipulate the metadata provided by the mass of users who take advantage of labour and money-saving web services (Srnicek 2017). This has supplanted the contract framework that had previously been underwritten by Hobbes' Leviathan state.

Attacks on traditional models of expertise may have been going on for some time, but they have been supercharged by the waves of digitalisation that emerged out of Silicon Valley (Weinberger 2011). The waves started in hippy culture during the 1970s but really took off in the early 1980s when hackers developed personal computers to give themselves the tools to manipulate information from existing knowledge infrastructures. Central among these was the futurologist Stewart Brand, who famously stated at the first Hackers' Conference in 1984 that 'information wants to be expensive, because it's so valuable' but that also 'information wants to be free, because the cost of getting it out is getting lower all the time. So you have these two fighting against each other.'[58]

It was not possible to square the apparent contradiction that information was free and at the same time made money to live on in the 1980s, but this was not necessarily driving the most radical of the hacking community. For these people, the central opportunity of a new information ecology was the creation of an enlightened politics from which a new mode of government might emerge, interlinked by microcomputers in cyberspace. The information revolution would refashion humans into netizens, who would in turn 'flatten organizations, globalize society, decentralize control, and help harmonize people' (Turner 2008). The ethos that this new digital generation worked towards was 'egalitarian, harmonious and free' (Turner 2008). The result would bring about a peaceful revolution that saw the end of the 'bureaucracies of the marketplace by stripping away the material bodies of the individuals and corporations' (Turner 2008). The new society would depend on flexible modes of production where everyone would have to be capable of working across multiple communities and domains in what would eventually be idealised as a network entrepreneur.

While entirely compatible with libertarian values, this ethos has a heritage that stretches right back to the hippies of the 1960s and ’70s and is defined by Richard Barbrook and Andy Cameron as ‘The Californian Ideology’, an ideology that has driven so much of the digital revolution over the past fifty years (see also Foer 2017).[59] Underpinning this ideology is the belief that self-government is better than government, the goal is to use a networked society to replace the power of the state. Indeed, it is quite possible to conceptualise Facebook as a part of the essential infrastructure of the city – like its rail, road and utilities infrastructure – one where people’s likes and engagements provide sufficient data to allow network managers to immediately address issues on a meta city-scale. While this benign interpretation of Facebook has lost much of its allure since the news that Cambridge Analytica was being used to micro-target political campaigns, the ‘Facebook Manifesto’ published by Mark Zuckerberg in a letter entitled ‘Building Global Community’ in 2017 can be ‘interpreted as an attempt to reassert the legitimacy of the projective city in the face of a global turn toward authoritarianism’ (Hoffman, Proferes and Zimmer 2016).

What makes datafication and the digital revolution properly radical, then, is the combination of attacks on existing categories of sense-making – including in the production of facts – from earlier eras and the potential to replace the Leviathan state with a digital platform. These changes are sucking in everything from labour relations to work, the switch from central banking to digital currency, what civil society can expect government to do, and how the armed forces fit into a dynamic, changing relationship with the state and society. Axiomatic categories like the objective world and subjective experience cannot do all the work that is demanded of them in the information infrastructures of the fourth dimension. Instead, we must refocus attention on the function of experience in social, professional and organisational relations and historicise the politics of the construction of experience as it relates to the technology platforms that frame our understanding and knowledge of war. Through this, we can start to see how emotional enclaves can be constructed and shaped to produce affective dissonance and solidarity in a digital environment that by definition disorders traditional social structures that depend on trust.

In this context, it becomes apparent that the armed forces are suffering from a digital dysphoria in which their central importance in the production of knowledge about war is being eroded through an ongoing process of digitalisation and datafication. In parallel with the broader assault on expertise, the new war ecology has enabled people to question professional opinion, flattening out the power of the expert while raising up the everyday voice of the amateur. This has created the perfect conditions for a breakdown in trust and the amplification of conspiracy theory that feeds into information operations. At the same time, by removing the bystander as a category in the hierarchy of violence, the number of targets and intelligence nodes has expanded beyond the traditional battlefield. The net result is radically reformulating the importance of the armed forces in the creation of knowledge about war even as political violence expands in ways that produce war of all against all.

CONCLUSION

Digitisation and the transformations it has stimulated have redefined the battlefield. These changes have been brought about through an immediate and ongoing interaction between connected technologies, human participants and the politics of violence. Contemporary information infrastructures ought to make it easier to know and make sense of war in the twenty-first century. In practice, the opacity of these systems renders these relationships more difficult to understand, locking digital individuals in informational prisms that can be hard to escape.

Under these new conditions, defining what counts as war is highly contextual. The Russian government, for example, has consistently denied that their armed forces have been involved in fighting in eastern Ukraine. While it has been obvious to Western intelligence agencies that Russia's military have been openly prosecuting a war against Ukraine,[1] British and American governments have chosen to ignore this reality, despite being signatories to the Budapest Memorandum that guaranteed Ukrainian sovereignty.[2] In October 2021, however, Aleksander Borodai, the newly elected Russian MP and former leader of the Donetsk and Luhansk People's Republics – the breakaway government of eastern Ukraine – contradicted the line taken by Moscow. Instead, Borodai stated

> of course, they're Russian people, Russian forces ... I think this is a crucial point. Here [i.e. Donbas] there were Russian forces,

> and there the armed forces of the Russian Federation – also Russian forces. Both the first and the second are Russian forces. Yes, one lot are called corps of the DPR and LPR people's militia, the others – corps of the Russian army. What's the difference?[3]

If Borodai is correct, then it seems Ukraine has been at war with Russia since 2014. However, Western powers have preferred to duck this reality. Instead, they struggled to respond to a war in which the aggressors were represented as 'little green men'. There are good political reasons why the West might prefer to view Russian involvement in Ukraine as something other than war. However, these reasons are easily entangled in political debates that undermine action or misdirect politics. So, for example, it is surprising to find that an online Russian disinformation campaign[4] was picked up by British politicians and used to drive home the incompetence of the European Union's foreign policy-making over Ukraine.[5] The tenor of the British argument was that the EU's attempt to draw Ukraine into closer alignment had provoked the Russians to support the rebellion in Donetsk and Luhansk. The net effect of the Russian information operation successfully sowed division within NATO and between the EU and the UK. In these circumstances, would NATO have been able to maintain that its defence of Ukraine was legitimate? Or would the social media feeds from the people of Luhansk and Donetsk signal that the West had already lost legitimacy even as NATO was dropping bombs in defence of Ukrainian sovereignty?

This constitutes the new war ecology, where mobile connected devices enable digital individuals to share and create content that influences politics and has lethal effects. The Russian government would not have had to do very much to demonstrate that ordinary citizens were being killed by Western powers. The citizens of Luhansk and Donetsk would have done all the propaganda work the Russians needed to justify further escalation. That the Ukrainian government could not control the information space to prevent news from emerging online reminds us that the new war ecology is distributed unevenly across the world. There are multiple media ecologies that are in various stages of development. Each of these new ecologies of war have their own dynamics, constraints, enablers and political, military and societal dynamics.

The underlying reasons for this shift in the way we think about war do not lie with the armed forces themselves but mirror an ongoing changing pattern of relationships between society and work. These have emerged through processes of digitalisation, in turn driving further globalisation. The process started in the late 1990s but has really taken off with the introduction of the first Apple iPhone and the first Android phone, in 2007 and 2008, respectively. This is creating all sorts of crises of representation that we have sought to highlight in this volume.

As we explained in Chapter 3, the trajectory and speed at which data emerges from the battlefield starts to become obvious once military information infrastructures are brought into contact with their civilian equivalents. In some parts of the world, highly connected societies render the battlefield transparent. In these contexts, people can upload events to YouTube or social media in real time. Armed forces must be prepared for narrative gaps to emerge as the stated reasons for a military operation are challenged by those who see inconsistency in the way rules of engagement are applied. These connected environments become subject to a range of counter-measures designed to shape both the tempo of battle and how war is represented online.

The smartphone has turned the MSM publication model on its head. Bound by legal constraints and editorial processes that demand stories are corroborated or fact-checked, traditional MSM struggles to keep up with the flood of news coming in from social media. Now people can produce, publish and consume media all from one device. They participate in war wherever they can get a wi-fi or network signal, irrespective of their immediate proximity to the fighting. And they do this at any time during the day, every day. This produces content that is out of synch with that of the MSM and it flattens civilian and military experiences into one register.

Even as they engage online, digital individuals must be careful about which networks they log in to and how different ISPs may be subject to hacking, disinformation and identity collection. Log in to the wrong network and you might be helping the enemy construct the target lists that they need for prosecuting the war. At the very least, a participant's online interactions may be subject to self-censorship in the hope that they do not attract attention from the wrong side.

Inevitably, this guarantees data spins off the battlefield and shapes the wider information environment in different ways. This leads to second-order online battlefields where different user groups try to make use of digital records for the purposes of building a case of war crimes or data is being pulled down by groups trying to tilt the news agenda. More fundamentally, the information infrastructures themselves go some way to shape what stories emerge, where they land and in what order and speed. This process reflects the heavily compartmentalised nature of government information infrastructures when compared to civilian ones. This also goes some way to explain how different narratives emerge in an official context and what happens to them as they meet discussions in more dynamic civilian social media environments.

To circumvent this churning of online narratives, the armed forces themselves are working on ways that help them accelerate warfare. As we saw in Chapter 3, this has two benefits. First, if decisions and operations occur at a faster tempo than those of the enemy, then there is the potential to defeat them in detail. Equally, if operations can be made to harness, rather than be subject to, the speed of flows of information and images about war, then there is the potential to control the online narrative. However, the prospect of realising such military and communication advantages must be seen in light of the existing poor state of government information infrastructures, where record-keeping, archival work and storing data has been dysfunctional. This has followed a twenty-year pattern of underinvestment and a lack of attention to how digitisation has undermined the bureaucracy's capacity to retrieve important corporate information. The result is twofold. First, the armed forces' capacity to produce enduring lessons that might help with improving performance is currently close to non-existent given the disjointed way in which the state collects, stores and makes the archive accessible. The second is that these failures are fatally undermining the methodological foundations of the historian.

As we saw in Chapter 4, the acceleration of warfare and the corresponding collapse of the archive creates further opportunity to shape post-trust social media prisms. Here online discussions combine and fold with established and more sedimented understandings of war, misrepresenting collective understanding in ways that constantly compete with and undermine expert voices. This has the effect of reframing how society makes sense of war, a process that is further complicated as transnational and global perspectives of war and remembrance are introduced. This points to how the schematisation of memory frames the way attention is being secured. As the archive becomes less reliable and politics more polarised, the space for developing a socially agreed perspective on the relationship between war and society reduces. The configuration of the internet and the nature of the new war ecology further encourages echo-chambers, information prisms and social media ghettos. This in turn guarantees the final triumph of memory over history, where not only is it harder to get access to digital archives but in which information ecologies also reward division and fragmentation.

This presents opportunities for creative media strategies that influence and reshape multiple narratives of war. At the same time, the dysphoria that these changes bring encourages an emotional desire for the security blanket of a sedimented history. This creates a sense of restorative nostalgia for a memory of the past that has been schematised for the purposes of managing national or digital memorial discourses. This has a powerful effect on shaping how people see themselves within society, framing attention and helping to shape political discourses.

As we discussed in Chapter 5, connected devices share data that can be weaponised and used for military purposes. Processes of datafication increase the datapoints available to those trying to identify patterns of life. These patterns of life are premediated through the technologies that are employed. For the targeteer, this facilitates a collapse in the distinction between willing and unwilling participants in war. The flood of datapoints and records that this creates helps put the archive to war. The digital archive thus becomes both an information repository and the means for target identification through data mining. This changes patterns of adversary identification from viewing enemies in terms of their iconic status to viewing them as part of an ongoing process of creating infinite targets.

Finally, Chapter 6 is concerned with how control and influence work in the new war ecology. Here we see the intersection of information infrastructures and the changing relationship between military actors and the private sector. Despite their ambition, defence establishments have struggled to develop and implement contemporary information infrastructures on their own. The relations between the government and service providers, technologists and the systems they are responsible for point towards a geometry of political power that has been inadequately surveyed. This is particularly problematic in the context of a new technology cold war between the United States and China. This is skewing public–private sector relationships, militarising wider society and further contributing to processes of datafication and participative surveillance. Only the virtual classes have the technical skills necessary for managing these activities. This in turn promotes an increased polarisation of technocratic skills away from government control.

The social and technological changes that these information infrastructures have unleashed are now having an effect on how the military organise and prepare for war. People have the means to participate in war in ways that have never previously been available. At the same time, digital devices are the basic sensors that the armed forces need for identifying targets and understanding patterns of life. In one device, participants have the means to record, consume and reproduce war. Silicon Valley has given the public a device that symbolises how far behind the military are in relation to shaping how contemporary war is fought.

Only from the 2010s have armed forces begun to comprehend, albeit without readily acknowledging this, how disconnected their way of working is from the rest of society. People in uniform know this. They know it takes longer to order items from their military supply chain than it takes Amazon to deliver the same thing more cheaply and more quickly to their office or barracks. The armed forces can see that their own approaches continue to be framed by well-established patterns of thinking. Yet the military prefer to try to bend existing martial cultures into this new and highly connected way of working.

Despite the many considerable efforts to the contrary, the armed forces continue to think about the world in twentieth-century terms. This reflects deeply established ways of working that are typically framed by analogue and highly sedimented depictions of war. For the armed forces, the goal is to hold on to existing bureaucratic and professional modes of organisation. There are air, land and sea domains. Cyber-warfare and information operations constitute new fields that must be added to these core organising structures. This ensures that each branch of the military can sustain and justify their existence as part of the whole. Doctrinally, these branches must be fused in such a way as to accelerate warfighting and try to prevent an adversary from exploiting any gaps that might otherwise result from this long-standing division of labour and framing of expertise.

While military doctrine points at how the armed forces see the world, it does not reflect how war comes to have meaning among society more broadly. The ramifications of these technologies and discourses in terms of how armed forces relate to wider society and the new war ecology have barely been touched upon. In a world with highly polarised perspectives, how armed forces sustain their apolitical relationship to a government bureaucracy that is itself working out how to deal with the fourth industrial revolution is not clear. There is a real danger that armed forces will seek to politicise their roles in an attempt to sustain themselves even as they undergo the disintermediating effects of digitalisation.

Over the course of this volume, we have sought to explore the implications of this shift in the representation and experience of war across the dimensions of data, attention and control. What we have found is that contemporary war is legitimised, planned, fought, experienced, remembered and forgotten in a continuous and connected way, through digitally saturated fields of perception. In *Radical War*, we have argued that war and its representation have collapsed into each other.

EPILOGUE

There are many ways in which Radical War polarises, distorts and undermines social and military cohesion. Digital individuals occupy the new war ecology because they get access to networked and connected services that give users more benefits than if they maintained only an analogue existence, even if that was possible. Social media platforms are designed to create a craving for hits and likes. In the process, they flatten out our appreciation for political violence, decontextualising it from its immediate social circumstances. This experience is then reframed through a global online community of like-minded followers. This leaves people polarised and isolated from wider communities.

The blowback from the wars in Iraq and Afghanistan, for example, are affecting Anglo-American politics in unexpected ways. Defeat has left some veterans wondering what these wars were about.[1] Some have concluded that metropolitan policy elites let the armed forces down. In August 2021, US Marine Corps Lieutenant Colonel Stuart Scheller took to Facebook to question the military and political chain of command and demand accountability for mistaken decisions in relation to Afghanistan.[2] Knowing that his videos would certainly damage his career, Scheller subsequently recorded a video for YouTube and declared: 'Follow me and we will bring the whole fucking system down.'[3] In the context of the 6 January 2021 insurrection at the Capitol Building, this invocation to his audience reflected his emotional response to events in Kabul but also implied a call for action. Some in the Marine Corps chain of command viewed this as an indication that Scheller wanted to see an insurrection in Washington, DC and Trump's restoration to the presidency.[4] Scheller denies that he wanted a revolution against the US constitution.

However, that has not stopped his online engagements from having political effect. In this case, Republican Congressman Louie Gohmert and Republican Congresswoman Marjorie Taylor Greene spoke in support of Scheller. Both legislators are pro-Trump, but they also have links to the QAnon conspiracy theory. The QAnon theory posits that 'Donald Trump is waging a secret war against elite Satan-worshipping paedophiles in government, business and the media.'[5] QAnon supporters were heavily represented in the insurrection at the Capitol Building on 6 January. At the same time, conspiracy theory has been brought into the heart of politics by those congressmen and women who take QAnon seriously.

Similarly, conspiracy theory is now a feature of British politics. In Britain's case, former members of the Parachute Regiment and veterans from Iraq and Afghanistan have been involved in COVID-19 anti-vaccine protests. Having sought to gain entry to old BBC studios in protest against MSM propagating what they consider to be pro-vaccine propaganda, one ex-soldier declared: 'Basically the men of our unit in our service, believe that we're pointing the weapons in the wrong direction.'[6] Recorded on a smartphone by an apparent member of the anti-vax political party the Freedom Alliance, the soldier when on to say:

> This time now the tyranny is against our people and we can't see it 'cos it's on our home soil where it's never been before. Because its psychological warfare not bombs, we can't see it, because its invisible. We've had this experience and used these tactics in other countries to manipulate, divide and conquer and now we're watching our own government and our own military use it against us. But the only men and women in this country that can resist against that are the ones that have the experience and the training that we use to help us.[7]

Social media has done a great deal to create the media ecosystems where people who share counter-cultural views can meet and organise. When presented as an affirmation of free speech, the danger is that the conspiracy theory becomes the reality not the exception. In many respects, the effects on political action are not always easy to see. However, there is a very strong possibility that online echo-chambers will lead to a further radicalisation of politics, where the tools and techniques applied overseas become the means by which social division is instrumentalised for political effect at home.

And yet it is also clear from the COVID-19 crisis that people can look past the effects of digitalisation on their relationship to work, war and politics. During lockdown, people came out on the balconies of their apartments and sang opera in solidarity with their neighbours. Society collectively celebrated the health workers they had entrusted to keep them alive as the world faced the pandemic. States could work together to find shared ways to leverage resources for shared benefit. It was possible to see beyond what was discordant and the sources of fear and doubt. In this book, we have not offered solutions to the predicaments created out of Radical War. Rather, we hope that by mapping the transformations of war in the twenty-first century we can contribute to a dialogue that reflects on and reconsiders the world as it is shaped by connected technologies, human participants and the politics of violence. We would like to be part of and join you in that debate at www.radicalwar.com.

APPENDIX 1

GLOSSARY OF TERMS

Archive

The idea of the archive has long been seen as the external and institutional basis for the remembering of societies at different stages of development across history and as an ultimate storage medium and metaphor of memory. It is seen as a reassuring and reliable repository, offering a secure place for the past's protection from loss, theft and erasure (Yeo 2017, p. ix).

In *Radical War*, however, we follow Moss and Thomas (2019, p. 117) in arguing that '[f]ar from being an object that is archived ... the internet is itself an archive, but one which does not conform to the rules of archiving as we know them'. They go on to argue that '[t]hose of us who work in the field of memory institutions need to confront this new world in which the internet is not archived but is the archive, not by claiming that it is not but by exploring its properties and possibilities'.

Datafication

Datafication is a process in which all aspects of life, including subjects, objects and practices, are being turned into online quantifiable datapoints (Southerton 2020).

Deep Mediatisation

According to Couldry and Hepp (2017), deep 'mediatization' represents a point in which every aspect and element of social life is composed of other elements that have been mediated. Datafication constitutes a good example of this, where lived experience and the material structures we depend on are all mediated by information technology and data-processing systems.

Digitisation versus Digitalisation

Digitisation is the process of encoding analogue information into zeroes and ones so that it can be stored, processed and transmitted by computers.[1]

According to the IT market analyst company Gartner, digitalisation is the process of using digital technologies to change business models and exploit the opportunities that emerge out of this activity.[2]

Digital Individual

According to Philip Agre (1994): 'The digital individual is the form of social identity that individuals acquire as their activities become influenced by – and often mediated through – digital representations of themselves.' The digital individual extends to the data footprints that are left when individuals engage with and fold into processes of participatory surveillance.

Fourth Dimension

The datafication of human experiences are intimately wrapped up into a new dimension sometimes called cyberspace. In this new space, the digital and the analogue can still be seen separately but where they also form a singular world (Scott 2015). Experiences in this fourth dimension take the notion of singularity for granted. Consequently, digital technologies disappear out of descriptions of this singular world even as they enable humans to participate in it.

Fourth Turning

The Fourth Turning is a book written by two historians (Strauss and Howe 1997). This book makes the case that world events unfold in a predictable cycle of around eighty years. Each cycle is made up of four turnings that are focused on growth, maturation, entropy and destruction. Mirroring some of the key ideas expressed in *The Fourth Turning*, key Donald Trump supporters like Steve Bannon have argued that American policy ought to be directed so as to 'get ahead of or stop any potential crisis'.[3] Their fear is that 'winter is coming'.

Hierarchy of Violence

Connected technologies and participatory surveillance establish a hierarchy of violence between those who actively and unwittingly participate. These hierarchies ensure that people facilitate violence irrespective of their political preferences or personal choices.

Information Infrastructures

Information infrastructures are a compound of people, processes, organisations and technical systems that include the services and technologies that make up the internet and process and transport data around the globe. These extended infrastructures take in the know-how and capabilities of the technologists who create, develop and build these systems of systems (Bowker et al. 2010).

Internet of Things

At its broadest sense, *Wired* magazine defined the IOT in 2018 as everything that is connected to the internet. More precisely, it can be defined as devices – from sensors to smartphones – that are connected together through the internet. These devices create and share data with each other, centrally via data centres and with specific application and hardware developers.[4] Developers may create their connected devices so that they can share information with each other, without sharing data centrally, or data may also be transferred to a central data warehouse.

Military Internet of Things

The MIOT can be defined as the multitude of sensors, weapons and devices that create and share data both with each other and with individual soldiers and headquarters depending on the cyber-security implications and the needs of commanders. The quantity of data that these systems produce will demand the introduction of different forms of algorithmic intervention so that appropriate metadata can be extracted, made sense of and then acted on (Kott, Swami and West 2016). This will create opportunities for those engaged in cyber-war.

New War Ecology

The new war ecology is a battle and information space that emerges out of war in the fourth dimension. It encompasses multiple media ecologies that are in various stages of development, depending on levels of connectivity to the internet, participation on social media, use of broadcast media and press freedoms.

Participative War

According to Merrin, participative war is a new mode of war 'where networked technologies and online public platforms allow anyone within or outside of a conflict zone to participate in informational war, to tell their story, expose events, offer support and contribute towards or expose propaganda' (Merrin 2018, p. 218).

Participatory Surveillance

A principal way in which we can speak of being a participant in digital living lies in the routine trade off in the exchange of information about ourselves (location, search history, identity, sexuality, contacts, personal relationships and so on) for access to a range of convenient services (including social media) and organisations. This trade off keeps digital individuals engaged in processes of participatory surveillance.

Planetary-scale Computation

Planetary-scale computation is made possible through the connection of software, hardware and networks across multiple, transnational layers (Bratton 2016) via fibreoptic data lines, communication satellites, cell phone networks and data centres. The architecture of planetary-scale computation is accidental in nature, emerging organically from people building it from below as much as by tech businesses creating it by design.

Post-trust

The processes of digitalisation driving a rise of the individual as centre of their own media worlds, a collapse in trust in the MSM and politics, and a loss of faith, certainty and security in history (Happer and Hoskins 2022).

The Radical Past

Our shared understanding of the past is caught between a pre-digital and highly sedimented appreciation for war in history as framed by analogue archives versus the digital churn of a present framed by social media. This agitation of history and memory is the radical past.

Schematisation

A schema is a framework or concept that helps us organise and interpret the world around us. These mental models represent shortcuts and standards that the mind forms from past experiences to help us understand and assimilate new experiences. The schematisation of war helps make sense of the velocity and volume of the billions of images of war that have now suddenly become available. The term has a long and influential history and is based on work by Frederic Bartlett (1932) and the neurologist Henry Head (1920), both of whom wrote about the psychology of memory.

Web 2.0

In 2005, *Wired* magazine defined Web 2.0 as a constellation of links between web applications that 'enabled users to generate content, rather than simply consume it … and … let developers get at the data'.[5] This made it possible to generate metadata about the specific tags and interests that motivated consumers. As a consequence, information could be free to consumers, and the data that their online activities produced could be used for the benefit of those trying to influence behaviours.

APPENDIX 2

KEY ICT INDICATORS

Source: ITU World Telecommunication/ICT Indicators Database

Key ICT indicators for developed and developing countries, the world and special regions (totals and penetration rates)

Active mobile-broadband subscriptions

	Millions															
	2005	*2006*	*2007*	*2008*	*2009*	*2010*	*2011*	*2012*	*2013*	*2014*	*2015*	*2016*	*2017*	*2018*	*2019*	*2020**
World	N/A	N/A	268	422	615	807	1,184	1,550	1,959	2,660	3,282	3,863	4,723	5,287	5,702	5,826
Developed	N/A	N/A	225	336	450	554	712	829	927	1,016	1,126	1,229	1,381	1,487	1,583	1,606
Developing	N/A	N/A	43	86	165	253	471	721	1,032	1,645	2,156	2,633	3,342	3,801	4,119	4,221
Least Developed Countries (LDCs)	N/A	N/A	0	1	1	3	11	25	42	95	141	192	258	292	329	351
Land Locked Developing Countries (LLDCs)	N/A	N/A	N/A	N/A	N/A	N/A	N/A	N/A	N/A	N/A	93	118	156	169	176	199
Small Island Developing States (SIDS)	N/A	N/A	N/A	N/A	N/A	N/A	N/A	N/A	N/A	N/A	22	26	31	34	38	39

Per 100 inhabitants

2005	*2006*	*2007*	*2008*	*2009*	*2010*	*2011*	*2012*	*2013*	*2014*	*2015*	*2016*	*2017*	*2018*	*2019*	*2020**
N/A	N/A	4.0	6.3	9.0	11.5	16.9	21.9	27.4	36.8	44.6	51.9	62.8	69.5	74.2	75.0
N/A	N/A	18.5	27.5	36.6	44.7	57.3	66.5	74.1	81.1	89.2	97.0	108.7	116.6	123.9	125.2
N/A	N/A	0.8	1.6	3.0	4.5	8.2	12.4	17.5	27.5	35.4	42.7	53.5	60.1	64.3	65.1
N/A	N/A	0.0	0.0	0.1	0.4	1.3	2.8	4.7	10.3	14.9	19.9	26.2	29.0	31.8	33.2
N/A	N/A	N/A	N/A	N/A	N/A	N/A	N/A	N/A	N/A	19.7	24.3	31.4	33.3	33.8	37.3
N/A	N/A	N/A	N/A	N/A	N/A	N/A	N/A	N/A	N/A	31.8	38.5	44.8	49.1	53.4	54.0

Population covered by a mobile-cellular network

	2005	*2006*	*2007*	*2008*	*2009*	*2010*	*2011*	*2012*	*2013*	*2014*	*2015*	*2016*	*2017*	*2018*	*2019*	*2020**
World	N/A	N/A	N/A	N/A	N/A	N/A	N/A	N/A	N/A	N/A	6,970	7,087	7,224	7,323	7,434	7,510
Developed	N/A	N/A	N/A	N/A	N/A	N/A	N/A	N/A	N/A	N/A	1,244	1,249	1,255	1,258	1,273	1,278
Developing	N/A	N/A	N/A	N/A	N/A	N/A	N/A	N/A	N/A	N/A	5,726	5,838	5,970	6,065	6,161	6,232
Least Developed Countries (LDCs)	N/A	N/A	N/A	N/A	N/A	N/A	N/A	N/A	N/A	N/A	810	839	860	888	914	940
Land Locked Developing Countries (LLDCs)	N/A	N/A	N/A	N/A	N/A	N/A	N/A	N/A	N/A	N/A	418	441	453	470	485	499
Small Island Developing States (SIDS)	N/A	N/A	N/A	N/A	N/A	N/A	N/A	N/A	N/A	N/A	61	60	62	62	63	64

Per 100 inhabitants

2005	*2006*	*2007*	*2008*	*2009*	*2010*	*2011*	*2012*	*2013*	*2014*	*2015*	*2016*	*2017*	*2018*	*2019*	*2020**
N/A	N/A	N/A	N/A	N/A	N/A	N/A	N/A	N/A	N/A	94.8	95.3	96.1	96.3	96.7	96.7
N/A	N/A	N/A	N/A	N/A	N/A	N/A	N/A	N/A	N/A	98.5	98.6	98.7	98.7	99.6	99.6
N/A	N/A	N/A	N/A	N/A	N/A	N/A	N/A	N/A	N/A	94.0	94.6	95.5	95.8	96.2	96.1
N/A	N/A	N/A	N/A	N/A	N/A	N/A	N/A	N/A	N/A	86.1	87.1	87.1	87.9	88.4	88.9
N/A	N/A	N/A	N/A	N/A	N/A	N/A	N/A	N/A	N/A	88.1	90.9	91.1	92.4	93.1	93.5
N/A	N/A	N/A	N/A	N/A	N/A	N/A	N/A	N/A	N/A	89.9	87.8	88.8	89.1	89.3	89.4

Population covered by at least a 3G mobile network

	2005	*2006*	*2007*	*2008*	*2009*	*2010*	*2011*	*2012*	*2013*	*2014*	*2015*	*2016*	*2017*	*2018*	*2019*	*2020**
World	N/A	N/A	N/A	N/A	N/A	N/A	N/A	N/A	N/A	N/A	5,756	6,280	6,610	6,900	7,128	7,235
Developed	N/A	N/A	N/A	N/A	N/A	N/A	N/A	N/A	N/A	N/A	1,188	1,219	1,226	1,231	1,249	1,254
Developing	N/A	N/A	N/A	N/A	N/A	N/A	N/A	N/A	N/A	N/A	4,569	5,061	5,384	5,669	5,879	5,981
Least Developed Countries (LDCs)	N/A	N/A	N/A	N/A	N/A	N/A	N/A	N/A	N/A	N/A	501	596	666	723	769	806
Land Locked Developing Countries (LLDCs)	N/A	N/A	N/A	N/A	N/A	N/A	N/A	N/A	N/A	N/A	236	282	322	350	376	400
Small Island Developing States (SIDS)	N/A	N/A	N/A	N/A	N/A	N/A	N/A	N/A	N/A	N/A	42	44	50	56	60	61

Per 100 inhabitants

2005	*2006*	*2007*	*2008*	*2009*	*2010*	*2011*	*2012*	*2013*	*2014*	*2015*	*2016*	*2017*	*2018*	*2019*	*2020**
N/A	N/A	N/A	N/A	N/A	N/A	N/A	N/A	N/A	N/A	78.3	84.5	87.9	90.8	92.8	93.1
N/A	N/A	N/A	N/A	N/A	N/A	N/A	N/A	N/A	N/A	94.0	96.2	96.5	96.6	97.7	97.7
N/A	N/A	N/A	N/A	N/A	N/A	N/A	N/A	N/A	N/A	75.0	82.0	86.2	89.6	91.8	92.2
N/A	N/A	N/A	N/A	N/A	N/A	N/A	N/A	N/A	N/A	53.3	61.9	67.5	71.6	74.4	76.2
N/A	N/A	N/A	N/A	N/A	N/A	N/A	N/A	N/A	N/A	49.8	58.1	64.7	68.8	72.2	74.9
N/A	N/A	N/A	N/A	N/A	N/A	N/A	N/A	N/A	N/A	61.5	63.4	72.7	80.0	85.5	85.7

Population covered by at least an LTE/WiMAX mobile network

	2005	*2006*	*2007*	*2008*	*2009*	*2010*	*2011*	*2012*	*2013*	*2014*	*2015*	*2016*	*2017*	*2018*	*2019*	*2020**
World	N/A	N/A	N/A	N/A	N/A	N/A	N/A	N/A	N/A	N/A	3,191	4,765	5,648	6,070	6,405	6,575
Developed	N/A	N/A	N/A	N/A	N/A	N/A	N/A	N/A	N/A	N/A	1,079	1,121	1,150	1,186	1,239	1,244
Developing	N/A	N/A	N/A	N/A	N/A	N/A	N/A	N/A	N/A	N/A	2,113	3,644	4,498	4,884	5,166	5,331
Least Developed Countries (LDCs)	N/A	N/A	N/A	N/A	N/A	N/A	N/A	N/A	N/A	N/A	145	188	232	333	387	429
Land Locked Developing Countries (LLDCs)	N/A	N/A	N/A	N/A	N/A	N/A	N/A	N/A	N/A	N/A	58	91	123	164	202	231
Small Island Developing States (SIDS)	N/A	N/A	N/A	N/A	N/A	N/A	N/A	N/A	N/A	N/A	24	28	36	39	43	44

Per 100 inhabitants

2005	*2006*	*2007*	*2008*	*2009*	*2010*	*2011*	*2012*	*2013*	*2014*	*2015*	*2016*	*2017*	*2018*	*2019*	*2020**
N/A	N/A	N/A	N/A	N/A	N/A	N/A	N/A	N/A	N/A	43.4	64.1	75.1	79.9	83.4	84.7
N/A	N/A	N/A	N/A	N/A	N/A	N/A	N/A	N/A	N/A	85.4	88.5	90.5	93.1	97.0	97.0
N/A	N/A	N/A	N/A	N/A	N/A	N/A	N/A	N/A	N/A	34.7	59.1	72.0	77.2	80.6	82.2
N/A	N/A	N/A	N/A	N/A	N/A	N/A	N/A	N/A	N/A	15.4	19.5	23.5	33.0	37.4	40.5
N/A	N/A	N/A	N/A	N/A	N/A	N/A	N/A	N/A	N/A	12.3	18.8	24.8	32.1	38.8	43.4
N/A	N/A	N/A	N/A	N/A	N/A	N/A	N/A	N/A	N/A	34.9	40.7	51.8	55.3	61.0	61.2

Key ICT indicators by urban/rural area (penetration rates)

Population covered by at least a 3G mobile network (%)

	Total						*Urban*										
	2015	*2016*	*2017*	*2018*	*2019*	*2020**	*2015*	*2016*	*2017*	*2018*	*2019*	*2020**	*2016*	*2017*	*2018*	*2019*	*2020**
World	78.3	84.5	87.9	90.8	92.8	93.1	96.6	98.2	99.2	99.4	99.6	99.4	68.2	74.3	80.1	84.2	84.8
Developed	94.0	96.2	96.5	96.6	97.7	97.7	98.4	100.0	100.0	100.0	100.0	100.0	82.4	83.5	84.0	89.3	89.1
Developing	75.0	82.0	86.2	89.6	91.8	92.2	96.0	97.7	99.0	99.2	99.4	99.2	67.0	73.5	79.8	83.8	84.5
Least Developed Countries (LDCs)	53.3	61.9	67.5	71.6	74.4	76.2	87.0	89.8	96.9	99.3	99.3	99.3	48.4	53.0	57.6	61.6	64.1
Land Locked Developing Countries (LLDCs)	49.8	58.1	64.7	68.8	72.2	74.9	85.7	93.1	97.8	100.4	99.7	99.7	42.9	50.2	54.7	59.8	63.7
Small Island Developing States (SIDS)	61.5	63.4	72.7	80.0	85.5	85.7	73.7	75.0	87.5	96.5	96.9	97.2	46.1	48.7	53.0	66.7	66.5

	2015	*2016*	*2017*	*2018*	*2019*	*2020**	*2015*	*2016*	*2017*	*2018*	*2019*	*2020**	*2016*	*2017*	*2018*	*2019*	*2020**
Africa	51.3	59.1	64.2	71.8	75.6	77.4	86.1	92.8	97.9	99.5	99.5	99.5	37.4	42.0	53.1	59.1	61.9
Arab States	74.6	83.3	86.9	90.1	90.7	90.8	92.5	99.7	99.8	99.8	99.8	99.8	63.3	71.2	76.2	77.5	77.7
Asia & Pacific	79.6	86.8	90.9	93.9	95.9	96.1	98.6	98.6	99.2	99.2	99.5	99.6	76.3	83.3	88.9	92.5	92.7
CIS	70.1	74.2	79.9	81.0	87.8	88.7	97.5	98.9	100.0	100.0	100.0	100.0	26.9	41.4	44.5	64.1	66.8
Europe	94.3	98.0	98.2	98.3	98.4	98.3	97.1	100.0	100.0	100.0	100.0	100.0	92.1	92.8	93.3	93.5	93.3
The Americas	91.3	92.8	94.0	94.3	95.4	95.5	97.8	98.5	99.2	99.4	99.4	98.2	69.5	72.0	72.5	78.2	78.0

Population covered by at least an LTE/WiMAX mobile network (%)

	Total						*Urban*										
	2015	*2016*	*2017*	*2018*	*2019*	*2020**	*2015*	*2016*	*2017*	*2018*	*2019*	*2020**	*2016*	*2017*	*2018*	*2019*	*2020**
World	43.4	64.1	75.1	79.9	83.4	84.7	64.1	81.7	89.8	92.2	94.3	95.1	43.2	57.3	64.8	69.7	71.4
Developed	85.4	88.5	90.5	93.1	97.0	97.0	92.3	94.2	95.1	97.7	100.0	100.0	67.7	73.8	76.0	85.8	85.6
Developing	34.7	59.1	72.0	77.2	80.6	82.2	54.8	77.7	88.2	90.4	92.6	93.7	41.0	55.8	63.8	68.4	70.2
Least Developed Countries (LDCs)	15.4	19.5	23.5	33.0	37.4	40.5	31.2	41.3	50.4	58.3	64.5	66.9	8.9	10.2	20.3	23.5	26.6
Land Locked Developing Countries (LLDCs)	12.3	18.8	24.8	32.1	38.8	43.4	35.7	51.4	63.9	72.8	82.3	84.4	4.6	7.6	14.1	19.3	24.7
Small Island Developing States (SIDS)	34.9	40.7	51.8	55.3	61.0	61.2	49.4	58.1	67.7	70.1	76.8	76.7	12.9	26.1	31.1	35.1	35.5

	2015	*2016*	*2017*	*2018*	*2019*	*2020**	*2015*	*2016*	*2017*	*2018*	*2019*	*2020**	*2016*	*2017*	*2018*	*2019*	*2020**
Africa	9.5	16.4	22.3	27.8	36.7	44.3	28.2	41.9	51.5	56.2	68.9	77.0	1.8	3.8	8.8	14.7	21.6
Arab States	18.6	27.1	49.9	60.2	61.9	61.9	34.6	44.7	71.4	76.1	75.9	75.7	3.8	21.2	39.6	43.9	43.9
Asia & Pacific	41.6	72.5	86.9	91.4	93.6	94.2	64.6	90.8	99.3	99.6	99.7	99.6	57.2	76.3	84.3	88.3	89.3
CIS	42.8	52.4	60.5	66.9	80.3	80.8	63.9	76.4	85.6	94.0	99.2	99.8	7.9	12.3	14.8	43.9	44.3
Europe	73.7	86.6	89.7	93.3	97.3	97.2	83.8	93.6	94.0	96.7	100.0	100.0	66.4	77.2	83.3	89.1	88.8
The Americas	72.9	77.7	81.8	86.2	88.5	88.7	84.2	88.8	91.8	95.8	97.0	98.4	36.9	39.3	47.8	53.9	54.1

NOTES

PROLOGUE

1. The total number can be calculated from looking at two Wikipedia entries. The first is periodised as 1990–2002. This page is available at: https://en.wikipedia.org/wiki/List_of_wars:_1990%E2%80%932002. The second is periodised 2003 to the present. This page is available at: https://en.wikipedia.org/wiki/List_of_wars:_2003%E2%80%93present. Both pages accessed 9 August 2020.
2. Moscow coined the term 'the near abroad' as a way of describing those republics that formed a Commonwealth of Independent States following the collapse of the Soviet Union. See 'Ethnic relations and Russia's "near-abroad"', *Encyclopaedia Britannica*. Available at: https://www.britannica.com/place/Russia/Ethnic-relations-and-Russias-near-abroad. Accessed 25 November 2021.
3. Wikipedia, Total Page Views. https://stats.wikimedia.org/#/all-projects. Accessed 30 September 2021.
4. 'The costs of war: human costs', The Watson Institute, Brown University, January 2020. Report available at: https://watson.brown.edu/costsofwar/costs/human. Accessed 9 August 2020.
5. ITU Statistics 2020. January 2020. Report available at: https://www.itu.int/en/ITU-D/Statistics/Pages/stat/default.aspx. Accessed 26 November 2021.
6. Internet Users, Internet Live Stats. Data correct as of 26 November 2021. Available at: https://www.internetlivestats.com/internet-users/. Accessed 26 November 2021.
7. ICT data for the world, by geographic regions, by urban/rural area

and by level of development for the following indicators (2005-2020; excel). Report available at: https://www.itu.int/en/ITU-D/Statistics/Documents/facts/ITU_regional_global_Key_ICT_indicator_aggregates_Nov_2020.xlsx. The ICT's main statistics page is available here: https://www.itu.int/en/ITU-D/Statistics/Pages/stat/default.aspx. Both pages accessed on 26 November 2021.

8. Ibid.
9. Number of smartphone users from 2016 to 2021 (in billions). *Statista*, 6 August 2021. Available at: https://www.statista.com/statistics/330695/number-of-smartphone-users-worldwide/. Accessed 26 November 2021.
10. Smartphones sold today, Internet Live Stats. Data correct on 25 November 2021. Available at: https://www.internetlivestats.com/watch/smartphones-sold/. Accessed 26 November 2021.
11. Ibid. ICT data for the world, by geographic regions, by urban/rural area and by level of development for the following indicators (2005-2020; excel).

INTRODUCTION

1. World Economic Forum Annual Meeting 2016: Mastering the Fourth Industrial Revolution, World Economic Forum, 2 February 2016. Report available at: https://www.weforum.org/reports/world-economic-forum-annual-meeting-2016-mastering-the-fourth-industrial-revolution. Accessed 13 August 2020.
2. See lecture given by Steve Blank, 'A secret history of Silicon Valley', Computer History Museum, 20 November 2008. Lecture can be found at: https://www.youtube.com/watch?v=ZTC_RxWN_xo. Accessed 13 August 2020.
3. Alex Hern, 'Smartphone is now "the place where we live", anthropologists say', *The Guardian*, 10 May 2021. Available at: https://www.theguardian.com/technology/2021/may/10/smartphone-is-now-the-place-where-we-live-anthropologists-say. Accessed 18 October 2021.
4. 'Reading stabbings: Libyan refugee Khairi Saadallah named as suspect in Forbury Gardens attack', Sky News, 22 June 2020. Available at: https://news.sky.com/story/reading-stabbings-khairi-saadallah-named-as-suspect-in-forbury-gardens-attack-12011810. Accessed 23 June 2020.
5. 'Man wearing "hoax explosive" shot dead in London Bridge attack', Al Jazeera, 30 November 2019. Available at: https://www.aljazeera.

com/news/2019/11/uk-met-police-confirm-stabbing-attack-london-bridge-191129150249684.html. Accessed 23 June 2020.
6. Statement of Dr Eric Schmidt, House Armed Services Committee, April 17, 2018. Statement available at: https://docs.house.gov/meetings/AS/AS00/20180417/108132/HHRG-115-AS00-Wstate-SchmidtE-20180417.pdf. Accessed 13 April 2020.
7. Dan Whitcomb, 'Former U.S. president Donald Tump launches "TRUTH" social media platform', Reuters, 21 October 2021. Available at: https://www.reuters.com/world/us/former-us-president-donald-trump-launches-new-social-media-platform-2021-10-21. Accessed 26 October 2021.
8. Mike Isaac and Damien Cave, 'Facebook strikes deal to restore news sharing in Australia', *New York Times*, 22 February 2021. Available at: https://www.nytimes.com/2021/02/22/technology/facebook-australia-news.html. Matthew Chapman, 'Revealed: Mark Zuckerberg threatened to pull UK investment in secret meeting with Matt Hancock', The Bureau of Investigative Journalism, 8 December 2020. Available at: https://www.thebureauinvestigates.com/stories/2020-12-08/revealed-mark-zuckerberg-threatened-to-pull-uk-investment-in-secret-meeting-with-matt-hancock. Both articles accessed 10 March 2021.

1. WAR AND THE DEMOCRATISATION OF PERCEPTION

1. 'Christchurch shooting: gunman Tarrant wanted to kill "as many as possible"', BBC News, 24 August 2020. Available at: https://www.bbc.co.uk/news/world-asia-53861456. Accessed 28 August 2020.
2. Kevin Roose, 'A mass murder of, and for, the internet', *New York Times*, 15 March 2019. Available at: https://www.nytimes.com/2019/03/15/technology/facebook-youtube-christchurch-shooting.html. Accessed: 30 September 2019.
3. Kjellberg has almost 90 million subscribers on YouTube. See E. J. Dickson, 'Why did the Christchurch shooter name-drop YouTube phenom PewDiePie?', *Rolling Stone*, 15 March 2019. Available at: https://www.rollingstone.com/culture/culture-news/pewdiepie-new-zealand-mosque-shooting-youtube-808633. Accessed 16 December 2019.
4. Facebook Newsroom tweet. Posted 16 March 2019.
5. Craig Timberg et al., 'The New Zealand shooting shows how YouTube and Facebook spread hate and violent images – yet again', *The Washington Post*, 16 March 2019. Available at: https://www.

washingtonpost.com/technology/2019/03/15/facebook-youtube-twitter-amplified-video-christchurch-mosque-shooting. Accessed: 30 September 2019.

6. Ibid.
7. 'Germany shooting: data on online spread of livestreamed attack kept secret', *The Guardian*, 19 October 2019. Available at: https://www.theguardian.com/media/2019/oct/19/germany-shooting-data-on-online-spread-of-livestreamed-attack-kept-secret. Accessed 17 December 2019.
8. Julia Carrie Wong, 'Germany shooting suspect livestreamed attempted attack on synagogue', *The Guardian*, 10 October 2019. Available at: https://www.theguardian.com/world/2019/oct/09/germany-shooting-synagogue-halle-livestreamed. Accessed 16 December 2019.
9. James Verini, 'How the battle of Mosul was waged on WhatsApp', *The Guardian*, 28 September 2019. Available at: https://www.theguardian.com/world/2019/sep/28/battle-of-mosul-waged-on-whatsapp-james-verini. Accessed 23 October 2021.
10. Shawn Snow, Kyle Rempfer and Meghann Myers, 'Deployed 82nd airborne unit told to use these encrypted messaging apps on government cell phones', *Military Times*, 23 January 2020. Available at: https://www.militarytimes.com/flashpoints/2020/01/23/deployed-82nd-airborne-unit-told-to-use-these-encrypted-messaging-apps-on-government-cellphones. Blake Moore and Jan E. Tighe, 'Insecure communications like WhatsApp are putting U.S. national security at risk', 8 December 2020. Available at: https://www.nextgov.com/ideas/2020/12/insecure-communications-whatsapp-are-putting-us-national-security-risk/170577. Both articles accessed 23 October 2021.
11. Stephanie Kirchgaessner, 'How NSO became the company whose software can spy on the world', *The Guardian*, 23 July 2021. Available at: https://www.theguardian.com/news/2021/jul/23/how-nso-became-the-company-whose-software-can-spy-on-the-world. Accessed 23 October 2021.
12. Dave Lee, 'WhatsApp discovers "targeted" surveillance attack', BBC News, 14 May 2019. Available at: https://www.bbc.co.uk/news/technology-48262681. Accessed 23 October 2021.
13. 'Exposed: Russia's military intel behind "destructive" Telegram channels prominent in Ukraine', UNIAN Information Agency, 1 February 2021. Available at: https://www.unian.info/society/infowars-russia-s-military-intel-behind-destructive-telegram-

channels-prominent-in-ukraine-11305508.html. Accessed 31 October 2021.

14. Stephen Losey, 'Woman shot and killed at Capitol was security forces airman, QAnon adherent', *Air Force Times*, 7 January 2021. Available at: https://www.airforcetimes.com/news/your-air-force/2021/01/07/woman-shot-and-killed-at-capitol-was-security-forces-airman-qanon-adherent. Accessed 30 October 2021.
15. Chris Stokel-Walker, 'Afghans are racing to erase their online lives', *Wired*, 17 August 2021. Available at: https://www.wired.co.uk/article/afghanistan-social-media-delete. Accessed 28 October 2021.
16. For a full analysis of the entire missile, drone and rocket artillery inventory for both Armenia and Azerbaijan, see Shaan Shaikh and Wes Rumbaugh, 'The air and missile war in Nagorno Karabakh: lessons for the future of strike and defense', Center for Strategic & International Studies, 8 December 2020. Available at: https://www.csis.org/analysis/air-and-missile-war-nagorno-karabakh-lessons-future-strike-and-defense. Accessed 28 October 2020.
17. Robert Bateman, 'No drones haven't made tanks obsolete', *Foreign Policy*, 15 October 2020. Available at: https://foreignpolicy.com/2020/10/15/drones-tanks-obsolete-nagorno-karabakh-azerbaijan-armenia. Seth J. Frantzman, 'How Azerbaijan's drones show what the future of war looks like', *Newsweek*, 7 October 2020. Available at: https://www.newsweek.com/how-azerbaijans-drones-show-what-future-war-looks-like-opinion-1536487. Both articles accessed 28 October 2021.
18. Matthew Gault, 'Azerbaijan dropped a music video before going to war with Armenia', *Wired*, 22 October 2020. Available at: https://www.vice.com/en/article/epdgjn/azerbaijan-dropped-a-music-video-before-going-to-war-with-armenia. Accessed 28 October 2021.
19. 'Freedom on the net 2021', country list available at: https://freedomhouse.org/countries/freedom-net/scores. Accessed 28 October 2021.
20. 'Freedom on the net 2021, Azerbaijan', Freedom House. Available at: https://freedomhouse.org/country/azerbaijan/freedom-net/2020. Accessed 28 October 2021.
21. 'Freedom on the net 2021, Armenia', Freedom House. Available at: https://freedomhouse.org/country/armenia/freedom-net/2021. Accessed 28 October 2021.
22. Christopher Paul and Miriam Matthews, 'The Russian "firehose of falsehood" propaganda model', RAND Corporation, 2016. Available at:

https://www.rand.org/pubs/perspectives/PE198.html. Accessed 12 November 2019.

23. Ben Norton, 'Behind NATO's "cognitive warfare": "battle for your brain" waged by Western militaries', MRonline, 13 October 2021. Available at: https://mronline.org/2021/10/13/behind-natos-cognitive-warfare-battle-for-your-brain-waged-by-western-militaries. Accessed 24 October 2021.
24. Andrew Radin, Alyssa Demus and Krystyna Marcinek, 'Understanding Russian subversion: patterns, threats, responses', RAND Corporation, 2020. Available at: https://www.rand.org/pubs/perspectives/PE331.html. Accessed 23 October 2021.
25. See, for example, Chris Tuck, 'What is Multi-Domain Integration?', Defence-in-Depth, 14 May 2021. Available at: https://defenceindepth.co/2021/05/14/what-is-multi-domain-integration. Accessed 24 October 2021.
26. Joint Concept Note (JCN 1/20), 'Multi-Domain Integration', Development Concepts and Doctrine Centre, UK Ministry of Defence. Available at: https://www.gov.uk/government/publications/multi-domain-integration-jcn-120. Accessed 28 October 2021.
27. 'Number of monthly active Facebook users worldwide as of 3rd quarter 2021', Statista Research Department, 1 November 2021. Available at: https://www.statista.com/statistics/264810/number-of-monthly-active-facebook-users-worldwide/. Accessed 25 November 2021.
28. More details about internet usage can be found here: https://twitter.com/ValaAfshar/status/1294843351603335168?s=20. Accessed 19 August 2020.
29. 'Frances Haugen says Facebook is "making hate worse"', BBC News, 26 October 2021. Available at: https://www.bbc.co.uk/news/technology-59038506. Accessed 1 November 2021.
30. Zak Doffman, 'Coronavirus reality check: yes, U.S. and Europe will track our phone location data – get used to it', *Forbes*, 19 March 2020. Available at: https://www.forbes.com/sites/zakdoffman/2020/03/19/coronavirus-reality-check-yes-us-and-eu-governments-will-track-our-phones-get-used-to-it. Accessed 16 April 2020.
31. Paul Mozur, 'Coronavirus outrage spurs China's internet police to action', *New York Times*, 16 March 2020. Available at: https://www.nytimes.com/2020/03/16/business/china-coronavirus-internet-police.html. Accessed 16 April 2020.
32. Zak Doffman, 'Coronavirus spy apps: Israel joins Iran and China tracking citizens' smartphones to fight COVID-19', *Forbes*, 14

March 2020. Available at: https://www.forbes.com/sites/zakdoffman/2020/03/14/coronavirus-spy-apps-israel-joins-iran-and-china-tracking-citizens-smartphones-to-fight-covid-19/#3d3fafa781bb. Accessed 16 April 2020.

33. Alex Hern, 'Experts warn of privacy risk as U.S. uses GPS to fight coronavirus', *The Guardian*, 2 April 2020. Available at: https://www.theguardian.com/technology/2020/apr/02/experts-warn-of-privacy-risk-as-us-uses-gps-to-fight-coronavirus-spread. Accessed 16 April 2020.
34. Ibid.
35. Ishaan Tharoor, 'Viewpoint: why was the biggest protest in world history ignored?', *Time Magazine*, 15 February 2013. Available at: https://world.time.com/2013/02/15/viewpoint-why-was-the-biggest-protest-in-world-history-ignored. Accessed 19 August 2020.
36. Alex Callinicos, 'Anti-war protests do make a difference', *Socialist Worker* online, 19 March 2005. Available at: https://web.archive.org/web/20060321084247/http://www.socialistworker.co.uk/article.php?article_id=6067. Accessed 19 August 2020.
37. The population of Iraq in 2003 was approximately 25 million.
38. 'Text of Bush speech', CBS News, 1 May 2003. Text available at: https://www.cbsnews.com/news/text-of-bush-speech-01-05-2003. Accessed 4 August 2020.
39. *Once Upon a Time in Iraq*, BBC documentary series, first broadcast 15 July 2020. Programme website available at: https://www.bbc.co.uk/programmes/m000kxws. Accessed 19 August 2020.
40. For example, opinion poll findings in 2014 found that 37 per cent of the British public said that Britain 'should stop trying to protect international influence and just concentrate on issues at home'. And by November 2018, 52 per cent of UK adults opposed military intervention overseas. For a more detailed breakdown of the British public's unwillingness to get involved in wars overseas, see Flora Holmes, 'Public attitudes to military interventionism', The British Foreign Policy Group, January 2020. Available at: https://bfpg.co.uk/2020/01/public-attitudes-to-uk-military-interventionism. Accessed 8 July 2020.
41. The two hits for the range of this five-hour to two-year spread of search results were: 'The war in Syria: regime attacks target civilians in Idlib' by TRT World and Syrian Blood Compilation Part 2 by 'War Leaks'. YouTube search made on 6 August 2019.
42. Richard Hall and Borzou Darangahi, 'Doctors in Idlib will no longer share coordinates of hospitals with UN after repeated attacks from

Russian Syrian Forces', *The Independent*, 3 June 2019. Available at: https://www.independent.co.uk/news/world/middle-east/syria-hospital-bombings-idlib-un-doctors-russia-assad-attack-a8942076.html. Accessed 25 September 2019.

43. Airwars Report, 'News in brief: U.S. media coverage of civilian harm in the war against so-called Islamic State', Airwars, 16 July 2019, p. 44. Available at: https://airwars.org/wp-content/uploads/2019/07/Airwars-News-in-Brief-US-media-reporting-of-civilian-harm.pdf. Accessed 16 August 2019.
44. Ibid., p. 40.
45. Ryan Browne, 'Pentagon report calls for changes to how the U.S. measures civilian deaths', CNN, 6 February 2019. Available at: https://edition.cnn.com/2019/02/05/politics/pentagon-civilian-deaths/index.html. Accessed 20 August 2019.
46. 'U.S.-led coalition in Iraq and Syria', Airwars. Available at: https://airwars.org/conflict/coalition-in-iraq-and-syria. Accessed 22 August 2019.
47. Arron Merat, '"The Saudis couldn't do it without us": the UK's true role in Yemen's deadly war', *The Guardian*, 18 June 2019. Available at: https://www.theguardian.com/world/2019/jun/18/the-saudis-couldnt-do-it-without-us-the-uks-true-role-in-yemens-deadly-war. Accessed 16 December 2019.
48. 'RAF killed "4,000 fighters in Iraq and Syria"', BBC, 7 March 2019. Available at: https://www.bbc.co.uk/news/uk-47477197. Accessed 16 December 2019. See earlier references to Airwars' open-source data analysis for an explanation of why this casualty figure is dubious. This BBC article is itself an example of how the media avoid challenging state narratives by hiding the one civilian death in the main body of the piece rather than in the by-line where it will be picked up on social media.
49. Discussion with Chris Woods, director of Airwars, 18 June 2019.
50. Andrew Dyer, 'In a tweet, Trump tells Navy not to boot Gallagher from SEALs', *San Diego Union-Tribune*, 21 November 2019. Available at: https://www.sandiegouniontribune.com/news/military/story/2019-11-21/in-a-tweet-trump-tells-navy-not-to-boot-gallagher-from-the-seals. Accessed 25 November 2021.
51. Dave Philipps, 'Navy SEALs were warned against reporting their chief for war crimes', *New York Times*, 23 April 2019. Available at: https://www.nytimes.com/2019/04/23/us/navy-seals-crimes-of-war.html. Accessed 16 December 2019.
52. Andrew Hoskins, 'The radicalisation of memory: monuments and

memorials in a post-trust era', keynote talk, Moving Monuments Conference, Manchester Centre for Public History & Heritage, Manchester Metropolitan University, 20 April 2018.

53. John Spencer, 'How the military is making it hard to remember our wars', *The Washington Post*, 10 November 2017. Available at: https://www.washingtonpost.com/outlook/how-the-military-is-making-it-hard-to-remember-our-wars/2017/11/10/ff7d6d4e-c324-11e7-aae0-cb18a8c29c65_story.html?noredirect=on. Accessed 19 August 2019.
54. 'Facebook admits it was used to "incite offline violence" in Myanmar', BBC News, 6 November 2018. Available at: https://www.bbc.co.uk/news/world-asia-46105934. Accessed 20 August 2020.
55. 'Human rights impact assessment: Facebook in Myanmar', BSR 2018. Available at: https://www.bsr.org/en/our-insights/blog-view/facebook-in-myanmar-human-rights-impact-assessment. Accessed 30 October 2020. Facebook commissioned this BSR report and accepted its findings in November 2018.
56. Ibid.
57. UNHRC, thirty-ninth session, 'Report of the detailed findings of the independent international fact-finding mission on Myanmar', 17 September 2018, p. 326. Available at: https://www.ohchr.org/Documents/HRBodies/HRCouncil/FFM-Myanmar/A_HRC_39_CRP.2.pdf. Accessed 16 August 2019.
58. Aung Naing Soe cited in Timothy McLaughlin, 'How Facebook's rise fueled chaos and confusion in Myanmar', *Wired*, 6 July 2018. Available at: https://www.wired.com/story/how-facebooks-rise-fueled-chaos-and-confusion-in-myanmar. Accessed 16 August 2019.
59. 'Frances Haugen says Facebook is "making hate worse"', BBC News, 26 October 2021. Available at: https://www.bbc.co.uk/news/technology-59038506. Accessed 1 November 2021.
60. UNHRC, thirty-ninth session, p. 340.
61. Ibid., pp. 339–44.
62. Ibid., p. 429.
63. Amanda Taub and Max Fisher, 'Where countries are tinderboxes and Facebook is a match', *New York Times*, 21 April 2018. Available at: https://www.nytimes.com/2018/04/21/world/asia/facebook-sri-lanka-riots.html?module=inline. Accessed 16 August 2019.
64. Max Fisher, 'Sri Lanka blocks social media, fearing more violence', *New York Times*, 21 April 2019. Available at: https://www.nytimes.com/2019/04/21/world/asia/sri-lanka-social-media.html?module=inline. Accessed 16 August 2019.

65. 'Sri Lanka: Facebook apologises for role in 2018 anti-Muslim riots', Al Jazeera, 13 May 2020. Available at: https://www.aljazeera.com/news/2020/5/13/sri-lanka-facebook-apologises-for-role-in-2018-anti-muslim-riots. Accessed 25 November 2021.
66. Cited in: Airwars Report, 'News in brief: U.S. media coverage of civilian harm in the war against so-called Islamic State', p. 39.
67. Angel Rabasa et al., 'The lessons of Mumbai', RAND Corporation, Santa Monica, 2009. Report available at: https://www.rand.org/pubs/occasional_papers/OP249.html. Accessed 21 August 2020.
68. 'Paris attacks: what happened on the night', BBC News, 9 December 2015. Available at: https://www.bbc.co.uk/news/world-europe-34818994. Accessed 21 August 2020.
69. Rukmini Callimachi, Alissa J. Rubin and Laure Fourquet, 'A view of ISIS's evolution in new details of Paris attacks', *New York Times*, 19 March 2016. Article available at: https://www.nytimes.com/2016/03/20/world/europe/a-view-of-isiss-evolution-in-new-details-of-paris-attacks.html. Accessed 21 August 2020.
70. Ibid.
71. Franco 'Bifo' Berardi, 'The coming global civil war: is there any way out?', e-flux Journal 69 (January 2016). Available at: https://www.e-flux.com/journal/69/60582/the-coming-global-civil-war-is-there-any-way-out. Accessed 16 December 2019.

2. UNDERSTANDING THE NEW WAR ECOLOGY

1. Andrew Hoskins giving evidence to the Airspace Tribunal London Hearing, Doughty Street Chambers, 21 September 2018. Available at: http://airspacetribunal.org/about/london-hearing. Accessed 4 November 2021.
2. Benjamin Jensen and John Paschkewitz, 'Mosaic warfare: small and scalable are beautiful', War on the Rocks, 23 December 2019. Available at: https://warontherocks.com/2019/12/mosaic-warfare-small-and-scalable-are-beautiful. Accessed 28 December 2019.
3. Gwen Ackerman, Selcan Hacaoglu and Mohammed Hatem, 'The drone wars are already here', Bloomberg Businessweek, 30 October 2019. Available at: https://www.bloomberg.com/news/articles/2019-10-30/the-drones-wars-are-here-and-they-re-escalating. Accessed 6 November 2019.
4. Dominic Cummings, 'Regime change #2: a plea to Silicon Valley – start a project NOW to write the plan for the next GOP candidate – The goal is not "reform" but a *government that

actually controls the government*', Dominic Cummings substack, 1 September 2021. Available at: https://dominiccummings.substack.com/p/regime-change-2-a-plea-to-silicon. Accessed 23 October 2021.

5. Jay Winter, 'Photographing war: archiving the visual record of war', Lecture given at the Archives of War: Media, Memory, History Conference, The National Archives, Kew, London, 30 November 2015.
6. Jeff Jarvis, 'Scorched earth', Medium, 9 February 2019. Available at: https://medium.com/whither-news/we-are-not-being-honest-with-ourselves-about-the-failures-of-the-models-we-depend-upon-803e491eda10. Accessed 5 November 2019.
7. Ibid.
8. '"Highway of death", Iraqi army armed retreat from Kuwait 1991', The History Channel. The History Channel does not keep an archive of its pre-internet shows available to search and so the only representations of this battle are the clips that have been cut and reproduced by YouTube users. Consequently, it is not possible to state the exact broadcast date of this programme. See, for example, https://youtu.be/hhmXleZXAr0. Accessed 11 November 2019.
9. Tony Blair's April 1999 Chicago Speech entitled, 'Doctrine of the international community' is available at: http://www.britishpoliticalspeech.org/speech-archive.htm?speech=279. Accessed 25 November 2021.
10. 'Andy Card recalls telling President Bush about 9/11 attacks', Remembering 9/11, MSNBC, 10 September 2016. Report available at: https://www.youtube.com/watch?v=7fs2duxjpE4. Accessed 7 August 2020.
11. The speech was given from the deck of the USS *Abraham Lincoln*, which was located off the coast of San Diego. Bush never actually used the phrase mission accomplished, but the words were plastered on a banner that had been strapped to the carrier's central island. 'Text of Bush speech', CBS News, 1 May 2003. Available at: https://www.cbsnews.com/news/text-of-bush-speech-01-05-2003. Accessed 4 August 2020.
12. For a summary of criticisms in relation to the legal framework employed in Iraq, watch 'Ghosts of Abu Ghraib', Home Box Office Films, 2006. Documentary available at: https://www.youtube.com/watch?v=1ucVvqOVwZI. Accessed 7 August 2020.
13. Dexter Filkins, 'The fall of the warrior king', *New York Times*, 23 October 2005. Available at: https://www.nytimes.com/2005/10/23/magazine/ the-fall-of-the-warrior-king.html. Accessed 5 August 2020.

14. David Rieff, 'Blueprint for a mess', *New York Times*, 2 November 2003. Available at: https://www.nytimes.com/2003/11/02/magazine/blueprint-for-a-mess.html. Accessed 5 August 2020.
15. Ibid.
16. David W. Brown, 'Michael Yon's war', *The Atlantic*, 1 June 2010. Available at: https://www.theatlantic.com/politics/archive/2010/06/michael-yons-war/57483. Accessed 5 August 2020.
17. 'Analysis of Russia's information campaign against Ukraine', NATO Stratcom Centre of Excellence, 2014. Available at: https://www.stratcomcoe.org/analysis-russias-information-campaign-against-ukraine. Accessed 12 November 2019.
18. This troll strategy is set out in detail in Paul and Matthews, 'The Russian "firehose of falsehood" propaganda model'.
19. Nadiya Romanenko et al., 'The troll network', 4 October 2016. Available at: http://texty.org.ua/d/fb-trolls/index_eng.html. Accessed 2 September 2020.
20. Kirill Meleshevich and Bret Schafer, 'Online information laundering: the role of social media', Alliance for Securing Democracy, Policy Brief no. 2, January 2018. Available at: https://securingdemocracy.gmfus.org/wp-content/uploads/2018/06/InfoLaundering_final-edited.pdf. Accessed 12 November 2019.
21. Natalia Lihachova, interviewed by Andrew Hoskins, Ukraine, 28 July 2016.
22. Christopher Allen, 'Who owns Ukraine's media?', Al Jazeera, 18 May 2016. Available at: https://www.aljazeera.com/indepth/features/2016/04/owns-ukraine-media-160405130121777.html. Accessed 2 September 2020.
23. Dan Milmo, 'Frances Haugen: I never wanted to be a whistleblower; but lives were in danger', *The Guardian*, 24 October 2021. Available at: https://www.theguardian.com/technology/2021/oct/24/frances-haugen-i-never-wanted-to-be-a-whistleblower-but-lives-were-in-danger. Accessed 31 October 2021.
24. While this model draws its inspiration from work on the Holocaust, our argument is not intended to stretch Holocaust debate away from that field's core domains.
25. 'Hollande Says "France is at War"', *New York Times*, 16 November 2015, http://www.nytimes.com/live/paris-attacks-live-updates/hollande-says-france-is-at-war.
26. Bernard-Henri Levy, 'War: Thinking the Unthinkable', http://www.huffingtonpost.com/bernardhenri-levy/war-thinking-the-unthinka_b_8590406.html.

27. Alex Hern, 'Fitness tracking app Strava gives away location of secret US army bases', *The Guardian*, 28 January 2018. Available at: https://www.theguardian.com/world/2018/jan/28/fitness-tracking-app-gives-away-location-of-secret-us-army-bases. Accessed 13 December 2019.
28. Eliot Higgins, 'How Bellingcat uncovered Russia's secret network of assassins', WIRED, 4 February 2021. Available at: https://www.wired.co.uk/article/russia-bellingcat-poison. Accessed 26 November 2021.
29. Aric Toler, 'Russia's "anti-selfie soldier law": greatest hits and implications', Bell¿ngcat, 20 February 2019. Available at: https://www.bellingcat.com/news/uk-and-europe/2019/02/20/russias-anti-selfie-soldier-law-greatest-hits-and-implications. Accessed 14 November 2019.
30. Lori Cameron, 'Internet of Things meets the military and battlefield: connecting gear and biometric wearables for an Internet of Military Things and Internet of Battlefield Things', Computing Edge, IEEE, March 2017. Available at: https://www.computer.org/publications/tech-news/research/internet-of-military-battlefield-things-iomt-iobt. Accessed 14 November 2019.

3. THE RUPTURED BATTLEFIELD

1. Felicia Schwartz, 'Israel begins ground operations against Hamas in Gaza', *Wall Street Journal*, 13 May 2021. Available at: https://www.wsj.com/articles/israel-steps-up-airstrikes-against-hamas-in-gaza-tries-to-contain-violence-at-home-11620900941?mod=hp_lead_pos5. Accessed 7 October 2021.
2. Julian Kossoff, 'Israel accused of tricking major news outlets into reporting fake Gaza invasion to lure Hamas fighters into tunnels that were targeted for massive airstrikes', Business Insider, 15 May 2021. Available at: https://www.businessinsider.com/gaza-israel-used-media-reports-lure-hamas-fighters-into-tunnels-2021-5?r=US&IR=T. Accessed 7 October 2021.
3. Ran Bar-Zik, 'Why the IDF told Israelis near Gaza to turn off their webcams', *Haaretz*, 20 May 2021. Available at: https://www.haaretz.com/israel-news/tech-news/.premium-why-the-idf-told-israelis-near-gaza-to-shut-their-webcams-1.9827201. Accessed 5 October 2021.
4. Spencer Robinson and Iain Overton, 'The targeting of high-rises in Gaza: an analysis of Israel's air strikes on tall buildings in 2021',

Action on Armed Violence, 30 September 2021. Available at: https://aoav.org.uk/2021/the-targeting-of-high-rises-in-gaza-an-analysis-of-israels-air-strikes-on-tall-buildings-in-2021. Accessed 5 October 2021.

5. 'The view from Tel Aviv', The Kicker – Columbia Journalism Review Podcast, 14 May 2021. Podcast available at: https://podcasts.google.com/feed/aHR0cHM6Ly9mZWVkcy5zb3VuZGNsb3VkLmNvbS91c2Vycy9zb3VuZGNsb3VkOnVzZXJzOjIwOTQ5MjU0OS9zb3VuZHMucnNz/episode/dGFnOnNvdW5kY2xvdWQsMjAxMDp0cmFja3MvMTA0ODc3MDUxNA?ep=14. Accessed 6 October 2021.
6. David Reinsel, John Gantz and John Rydning, 'The digitization of the world: from edge to core', Data Age 2025, International Data Corporation White Paper, November 2018. Report available at: https://www.seagate.com/gb/en/our-story/data-age-2025. Accessed 2 December 2019.
7. Iain Overton, 'Don't expect transparency from a government run by WhatsApp', *The Guardian*, 27 April 2021. Available at: https://www.theguardian.com/commentisfree/2021/apr/27/transparency-tory-government-whatsapp-lobbying-cronyism. Accessed 26 October 2021.
8. Arturo Munoz, 'U.S. military information operations in Afghanistan effectiveness of psychological operations 2001–2010', RAND Corporation – National Defense Research Institute, 2012. Available at: https://www.rand.org/pubs/monographs/MG1060.html. Accessed 2 September 2020.
9. Stanley McChrystal, 'It takes a network: the new frontline of modern warfare', *Foreign Policy*, 21 February 2011. Available at: https://foreignpolicy.com/2011/02/21/it-takes-a-network. Accessed 14 May 2020.
10. Sami Yousafzai, '10 years of Afghan war: how the Taliban go on', *Newsweek*, 2 October 2011. Available at: https://www.newsweek.com/10-years-afghan-war-how-taliban-go-68223. Accessed 26 February 2019.
11. Ibid.
12. Alex Strick van Linschoten and Felix Kuehn, 'Separating the Taliban from al-Qaeda: the cores of success in Afghanistan', New York: New York University Center on International Cooperation, 2011. Available at: https://cic.es.its.nyu.edu/sites/default/files/gregg_sep_tal_alqaeda.pdf. Accessed 28 October 2021.
13. Toby Whitmarsh and Arnel David, 'If you are not the first you are the last: gaining adaptive edge through prototype', War Room, 21 May

2019. Available at: https://warroom.armywarcollege.edu/articles/if-you-are-not-first-you-are-last-gaining-an-adaptive-edge-through-prototype-warfare. Accessed 6 August 2019.

14. 'Accelerated Warfare: futures statement for an army in motion', statement by the Australian chief of the army, Lieutenant General Rick Burr, 8 August 2018. Available at: https://www.army.gov.au/sites/default/files/publications/commanders_statement_army_in_motion_a4_u.pdf. Accessed 7 March 2019.
15. Cian O'Driscoll, 'Can wars no longer be won?', The Conversation, 2 December 2019. Available at: https://theconversation.com/can-wars-no-longer-be-won-126068?utm_medium=amptwitter&utm_source=twitter. Accessed 4 December 2019.
16. Christopher Paul and Miriam Matthews, 'The Russian "firehose of falsehood" propaganda model', RAND Corporation, 2016. Available at: https://www.rand.org/pubs/perspectives/PE198.html. Accessed 12 November 2019.
17. Of the 5.1 million security clearances issued by the US government, over 1 million were to contractors. See Brian Fung, '5.1 million Americans have security clearances: that's more than the entire population of Norway', *The Washington Post*, 24 March 2014. Available at: https://www.washingtonpost.com/news/the-switch/wp/2014/03/24/5-1-million-americans-have-security-clearances-thats-more-than-the-entire-population-of-norway/?arc404=true. A full list of significant cyber incidents since 2006 is detailed by the Center for Strategic & International Studies. For the report see: https://www.csis.org/programs/technology-policy-program/significant-cyber-incidents. Both the article and the report were accessed on 19 November 2019.
18. Helen Warrell and Nic Fildes, 'Amazon strikes deal with UK spy agencies to host top-secret material', *The Financial Times*, 25 October 2021. Available at: https://www.ft.com/content/74782def-1046-4ea5-b796-0802cfb90260. Accessed 27 October 2021.
19. This is particularly noticeable in non-military parts of government where the security risks associated with record-keeping are less onerous. See Stephen Bounds, 'Making government record-keeping work', Information & Data Manager, 21 June 2018. Available at: https://idm.net.au/article/0012049-making-government-record-keeping-work. Accessed 26 November 2019.
20. John Spencer, 'How the military is making it hard to remember our wars', *The Washington Post*, 10 November 2017.

21. National Audit Office, 'Improving the performance of major equipment contracts: Ministry of Defence', Report by the Comptroller and Auditor General. Session 2021, 24 June 2021, HC298, p. 47. Available at: https://www.nao.org.uk/report/improving-the-performance-of-major-equipment-contracts. Accessed 26 October 2021.
22. 'What is Discovery?' a description of the UK's National Archive search engine. The reference to more than 32 million descriptions was gathered on 26 November 2019. See https://discovery.nationalarchives.gov.uk for the latest update on record descriptions kept by TNA, UK. Accessed 4 November 2021.
23. Dan Spokojny, 'We are not capable of learning the lessons of Afghanistan', The Duck of Minerva, 19 October 2021. Available at: https://www.duckofminerva.com/2021/10/we-are-not-capable-of-learning-the-lessons-of-afghanistan.html. Accessed 26 October 2021.
24. Craig Whitlock, 'Afghanistan Papers: at war with the truth', *The Washington Post*, 9 December 2019. Available at: https://www.washingtonpost.com/graphics/2019/investigations/afghanistan-papers/afghanistan-war-confidential-documents. Accessed 19 December 2019.
25. Jonathan Schroden, 'There was no "secret war on the truth" in Afghanistan', War on the Rocks, 16 December 2019. Available at: https://warontherocks.com/2019/12/there-was-no-secret-war-on-the-truth-in-afghanistan. Accessed 19 December 2019.
26. Brigadier Ben Barry, 'Operations in Iraq: an analysis from a land perspective January 2005 – May 2009', UK Ministry of Defence, 29 November 2010. Published under FOI2016/07003/77396. Available at: https://www.gov.uk/government/uploads/system/uploads/attachment_data/file/557326/20160831-FOI07003_77396_Redacted.pdf. Accessed 19 December 2019.
27. Official report of the Iraq Inquiry: Section 152, p. 423, para. 86. Available at: http://www.iraqinquiry.org.uk/media/246651/the-report-of-the-iraq-inquiry_section-152.pdf. Accessed 1 January 2017.
28. Official website of the Iraq Inquiry: http://www.iraqinquiry.org.uk/media/95382/2010-07-21-Transcript-Irwin-Palmer-S3.pdf#search=lessons. Accessed 1 January 2017.
29. Official website of the Iraq Inquiry: http://www.iraqinquiry.org.uk/media/236689/2009-12-14-transcript-riley-wall-s2.pdf. Accessed 1 January 2017.

30. Sir William Gage, The Report of the Baha Mousa Inquiry, volume 1, 2011. Available at: https://www.gov.uk/government/uploads/system/uploads/attachment_data/file/279190/1452_i.pdf. Accessed 1 January 2017.
31. Ibid.
32. Ibid.
33. Sir Thayne Forbes, The Report of Al Sweady Inquiry, December 2014. Available at: https://www.gov.uk/government/uploads/system/uploads/attachment_data/file/388292/Volume_1_Al_Sweady_Inquiry.pdf. Accessed 1 January 2017.
34. Ibid.
35. This is reflected in the Sir Alex Allan 'Records Review' undertaken in August 2014. Available at: https://www.gov.uk/government/publications/records-review-by-sir-alex-allan. Accessed 27 November 2019.
36. 'Policies relating to the retention of information on deployed MOD IT systems'. Response to an FOI request regarding policies in force from 2004 relating to the retention of information on UK Ministry of Defence IT systems. Available at: https://www.gov.uk/government/publications/policies-relating-to-the-retention-of-information-on-deployed-mod-it-systems. Accessed 26 November 2019.
37. Press Association, 'British government and Army accused of covering up war crimes', *The Guardian*, 17 November 2019. Available at: https://www.theguardian.com/law/2019/nov/17/british-government-army-accused-covering-up-war-crimes-afghanistan-iraq. Accessed 3 December 2019.
38. Sharon Weinberger, 'Meet America's newest military giant: Amazon', *MIT Technology Review*, October–November 2019. Available at: https://www.technologyreview.com/s/614487/meet-americas-newest-military-giant-amazon. Accessed 3 December 2019.
39. Anonymous, UK Ministry of Defence workshop, February 2021.
40. Andrew Hoskins, 'Are we losing the history of warfare?', 2 February 2015. Available at: https://archivesofwar.gla.ac.uk/are-we-losing-the-history-of-warfare-by-andrew-hoskins. Accessed 4 November 2021.

4. THE RADICAL PAST

1. Peter Dizikes, 'Study: on Twitter, false news travels faster than true stories', MIT News, 8 March 2018. Available at: http://news.mit.edu/2018/study-twitter-false-news-travels-faster-true-stories-0308. Accessed 25 July 2020.

2. Ted Kotcheff speaking in *Erase and Forget* (2017). Dir. Andrea Luka Zimmermann.
3. Press Association, '"Major celebration", to mark Falklands War anniversary', *The Guardian*, 26 June 2006. Available at: https://www.theguardian.com/politics/2006/jun/26/immigrationpolicy.military. Accessed 10 October 2019.
4. Spencer, 'How the military is making it hard to remember our wars'.
5. Josephe DeLappe, Iraqimemorial.org – Commemorating Civilian Deaths. Available at: www.iraqimemorial.org. Accessed 10 December 2019.
6. Sonia Morland, 'Crimes against humanity: an exploration of genocide and ethnic violence', *Studies in Ethnicity and Nationalism*, 18 November 2011. Available at: http://senjournal.co.uk/2011/11/18/crimes-against-humanity-an-exploration-of-genocide-and-ethnic-violence. Accessed 10 October 2019.
7. 'David Blight discusses 9/11, part 3'. YouTube, 8 January 2010. Available at: http://www.youtube.com/watch?v=108DcwjA864. Accessed 10 October 2019. The Battle of Antietam of 1862 is the bloodiest single-day battle in American history with over 22,000 casualties.
8. See Michael Shaw, 'Big media sent 3 of my favorite war photographers to Afghanistan and what they brought back were the near-same medevac shots', Reading the Pictures, 16 January 2011. Available at: http://www.bagnewsnotes.com/2011/01/big-media-sent-3-of-my-favorite-war-photographers-to-afghanistan-and-what-they-brought-me-back-were-the-near-same-medevac-shots. Accessed 10 October 2019.
9. Ibid.
10. Ibid.
11. Simon Norfolk, 'Simon Norfolk in conversation with Andrew Hoskins', Open Eye Gallery, Liverpool, 3 May 2012.
12. Robin McKie and Vanessa Thorpe, 'Digital Domesday Book lasts 15 years not 1000', *The Guardian*, 3 March 2002. Available at: https://www.theguardian.com/uk/2002/mar/03/research.elearning. Accessed 11 August 2020.
13. 'BBC archives, wiped, missing and lost'. Available at: https://www.bbc.co.uk/archive/bbc-archives--wiped-missing-and-lost/z4nkvk7. Accessed 10 December 2019.
14. Ian Cobain, 'Foreign Office hoarding 1m historic files in secret archive', *The Guardian*, 18 October 2013. Available at: https://www.

theguardian.com/politics/2013/oct/18/foreign-office-historic-files-secret-archive. Accessed 9 July 2020.

15. Simon Jenkins, 'No more remembrance days: let's consign the 20th century to history', *The Guardian*, 9 November 2017. Available at: https://www.theguardian.com/commentisfree/2017/nov/09/no-more-remembrance-days-consign-20th-century-history. Accessed 10 October 2019.
16. Ibid.
17. Robert Draper, 'Toppling statues is a first step toward ending Confederate myths', *The National Geographic*, 2 July 2020. Available at: https://www.nationalgeographic.com/history/2020/07/toppling-statues-is-first-step-toward-ending-confederate-myths. Accessed 25 July 2020.
18. Elizabeth Day, '#BlackLivesMatter: the birth of a new civil rights movement', *The Guardian*, 19 July 2015. Available at: https://www.theguardian.com/world/2015/jul/19/blacklivesmatter-birth-civil-rights-movement. Accessed 25 July 2020.
19. Ben Jacobs and Warren Murray, 'Donald Trump under fire after failing to denounce Virginia white supremacists', *The Guardian*, 31 August 2017. Available at: https://www.theguardian.com/us-news/2017/aug/12/charlottesville-protest-trump-condemns-violence-many-sides. Accessed 4 November 2021.
20. American Historical Association Statement on Confederate Monuments, August 2017. Available at: https://www.historians.org/news-and-advocacy/aha-advocacy/aha-statement-on-confederate-monuments. Accessed 30 September 2019.
21. Ibid.
22. CNN Opinion, 'Historians: "defending history" is complicated in the US', CNN, 20 August 2017. Available at: https://edition.cnn.com/2017/08/19/opinions/historians-confederate-statues-opinion-roundup/index.html. Accessed 10 October 2019.
23. David Blight, 'The civil war lies on us like a sleeping dragon', *The Guardian*, 20 August 2017. Available at: https://www.theguardian.com/us-news/2017/aug/20/civil-war-american-history-trump. Accessed 4 November 2021.
24. Ibid.
25. Michael Martelle (ed.), 'Exploring the Russian social media campaign in Charlottesville', National Security Archive, 14 February 2019. Available at: https://nsarchive.gwu.edu/news/cyber-vault/2019-02-14/exploring-russian-social-media-campaign-charlottesville. Accessed 10 October 2019; see also Alex Heath,

'Facebook removed the event page for white nationalist "Unite the Right" rally in Charlottesville one day before it took place', Business Insider, 14 August 2017. Available at: https://www.businessinsider.com/facebook-removed-unite-the-right-charlottesville-rally-event-page-one-day-before-2017-8?r=U.S.&IR=T. Accessed 10 October 2019.

5. THE WEAPONISED ARCHIVE

1. Charlie Winter, 'Media jihad: the Islamic State's doctrine for information warfare', The International Centre for the Study of Radicalisation and Political Violence, King's College London, 2017. Available at: https://icsr.info/2017/02/13/icsr-report-media-jihad-islamic-states-doctrine-information-warfare. Accessed 17 August 2020.
2. Charlie Winter and Jade Parker, 'Virtual caliphate rebooted: the Islamic State's evolving online strategy', Lawfare, 7 January 2018. Available at: https://www.lawfareblog.com/virtual-caliphate-rebooted-islamic-states-evolving-online-strategy. Accessed 20 August 2020.
3. Greg Miller, 'As Islamic State loses territory, it seeks to survive online', *Washington Post*, 13 February 2017. Available at: https://www.washingtonpost.com/world/national-security/as-islamic-state-loses-territory-it-seeks-to-survive-online/2017/02/13/56356f08-f150-11e6-8d72-263470bf0401_story.html. Accessed 16 August 2020.
4. Charlie Winter, 'Totalitarianism 101: the Islamic State's offline propaganda strategy', Lawfare, 27 March 2016. Available at: https://www.lawfareblog.com/totalitarianism-101-islamicislamic-states-offline-propaganda-strategy. Accessed 17 August 2020.
5. Quote taken from Winter, 'Media Winter', 2017.
6. Agence France-Presse, 'YouTube investigates automatic deletion of comments criticising China Communist Party', *The Guardian*, 27 May 2020. Available at: https://www.theguardian.com/technology/2020/may/27/youtube-investigates-automatic-deletion-of-comments-criticising-china-communist-party?CMP=Share_iOSApp_Other. Accessed 27 May 2020.
7. American Historical Association Statement on Confederate Monuments, August 2017.
8. Geof Bowker, 'Just what are we archiving?', *Limn* 6, 'The Total Archive', March 2016. Available at: https://limn.it/articles/just-what-are-we-archiving. Accessed 30 September 2019.

9. See talk by Brigadier (ret.) Iain Harrison, strategic engagement director QinetiQ, 'How can armies modernise while remaining resilient?', RUSI Land Warfare Conference, 4 June 2019. Available at: https://twitter.com/RUSI_org/status/1135876090482892800?s=20. Accessed 8 October 2019. See also Toby Whitmarsh and Arnel David, 'If you are not first you are last: gaining an adaptive edge through prototype warfare', War Room, United States Army War College, 21 May 2019. Available at: https://warroom.armywarcollege.edu/articles/if-you-are-not-first-you-are-last-gaining-an-adaptive-edge-through-prototype-warfare. Accessed 8 October 2019.
10. Arthur Holland Michel, 'How rogue techies armed the predator, almost stopped 9/11, and accidentally invented remote war', *Wired*, 17 December 2015. Available at: https://www.wired.com/2015/12/how-rogue-techies-armed-the-predator-almost-stopped-911-and-accidentally-invented-remote-war. Accessed 30 September 2019.
11. Ibid.
12. Ibid.
13. Sharon Weinberger, 'Hollywood and hyper-surveillance: the incredible story of Gorgon Stare', *Nature – International Journal of Science*, 11 June 2019. Available at: https://www.nature.com/articles/d41586-019-01792-5. Accessed 8 October 2019.
14. John Marion, 'Wide-Area Motion Imagery systems: evolution, capabilities and mission sets', RUSI Defence Systems, 5 January 2017. Available at: https://rusi.devalps.eu/publication/rusi-defence-systems/ wide-area-motion-imagery-systems-evolution-capabilities-and-mission?page=159. Accessed 30 September 2019.
15. The Memoto Lifelogging Camera (later called 'Narrative') was founded in Sweden in 2012 but ultimately was never successful in the consumer marketplace. Their website is no longer available.
16. Chelsea Dobbins et al., 'Creating human digital memories with the aid of pervasive mobile devices', *Pervasive Mobile Computing*, published online 25 October 2013; eventually appeared in vol. 12, June 2014, pp. 160–79 and cited in Chris Baraniuk, 'Take it easy: make the fridge track all your snacking', *New Scientist*, 11 January 2014, p. 21. Available at: https://www.newscientist.com/article/mg22129514-600-lifelogging-even-your-home-appliances-could-do-it. Accessed 30 September 2019.
17. 'President George W. Bush's address to the Joint Session of the 107th Congress from 20 September 2011'. Selected Speeches of President George W. Bush, 2001–8. Available at: https://georgewbush-whitehouse.archives.gov. Accessed 26 July 2020.

18. Edward Wong, 'Americans demand a rethinking of the "Forever War"', *New York Times*, 2 February 2020. Available at: https://www.nytimes.com/2020/02/02/us/politics/trump-forever-war.html. Accessed 27 July 2020.
19. 'President George W. Bush's address to the Joint Session of the 107th Congress from 20 September 2011'.
20. Brian Walden, *Walden on Villains*. First broadcast on BBC Two on 30 April 1999. Available at: https://genome.ch.bbc.co.uk/fbc77129087248bd91a7e386f8b2c40b. Accessed 4 November 2021.
21. At the time of the war, Hugh Roberts worked for the International Crisis Group, which proposed a negotiated way out of the Libyan civil war. See Hugh Roberts, 'Who said Gaddafi had to go?', *London Review of Books* 33(22) (2011), pp. 8–18. Available at: https://www.lrb.co.uk/v33/n22/hugh-roberts/who-said-gaddafi-had-to-go. Accessed 30 September 2019.
22. Ibid.
23. Ibid.
24. Jonathan Mahler, 'What do we really know about Osama bin Laden's death?', *New York Times Magazine*, 15 October 2015. Available at: https://www.nytimes.com/2015/10/18/magazine/what-do-we-really-know-about-osama-bin-ladens-death.html. Accessed 30 September 2019.
25. Hassina Mechaï, '"Disastrous" Libyan intervention was France's Iraq War, says aid veteran', Middle East Eye, 10 March 2018. Available at: http://www.middleeasteye.net/news/libyan-conflict-was-frances-iraq-says-former-msf-chief-1346416981. Accessed 30 September 2019.
26. Peter Waldman, Lizette Chapman and Jordan Robertson, 'Palantir knows everything about you: Peter Thiel's data-mining company is using War on Terror tools to track American citizens; The scary thing? Palantir is desperate for new customers', Bloomberg, 19 April 2018. Available at: https://www.bloomberg.com/features/2018-palantir-peter-thiel. Accessed 8 October 2019.
27. Ibid.
28. See Amarnath Amarsingham and Marc-André Argentino, 'The QAnon conspiracy theory: a security threat in the making'?, *CTC Sentinel* 13(7) (July 2020). Available at: https://ctc.usma.edu/the-qanon-conspiracy-theory-a-security-threat-in-the-making. Accessed 25 August 2020; for how Trump endorsed and then retracted his endorsement of QAnon, see Amanda Holpuch, 'White House says Trump doesn't know of QAnon, despite his tacit endorsement', *The*

Guardian, 23 August 2020. Available at: https://www.theguardian.com/us-news/2020/aug/23/donald-trump-qanon-conspiracy-theory. Accessed 25 August 2020.

29. Jeremy W. Peters, 'Bannon's worldview: dissecting the message of "The Fourth Turning"', *New York Times*, 8 April 2017. Available at: https://www.nytimes.com/2017/04/08/us/politics/bannon-fourth-turning.html. Accessed 8 June 2020.
30. David Ignatius, 'Why Mattis and Millen toppled their bridge of silence', *The Washington Post*, 4 June 2020. Available at: https://www.washingtonpost.com/opinions/why-mattis-and-mullen-toppled-their-bridge-of-silence/2020/06/04/c71b8f58-a698-11ea-bb20-ebf0921f3bbd_story.html. Accessed 8 June 2020.
31. US Bureau of Labour Statistics, 'Unemployment rate rises to record high 14.7 percent in April 2020', 13 May 2020. Available at: https://www.bls.gov/opub/ted/2020/unemployment-rate-rises-to-record-high-14-point-7-percent-in-april-2020.htm. Accessed 9 June 2020. 'Unemployment rates during COVID-19 pandemic', Congressional Research Service, R46554, 20 August 2021. Available at: https://sgp.fas.org/crs/misc/R46554.pdf. Accessed 26 November 2021.
32. Naomi Klein, 'Screen New Deal: under cover of mass death, Andrew Cuomo calls in the billionaires to build a high-tech dystopia', The Intercept, 8 May 2020. Available at: https://theintercept.com/2020/05/08/andrew-cuomo-eric-schmidt-coronavirus-tech-shock-doctrine. Accessed 9 June 2020.
33. Ibid.

6. TECHNOLOGIES OF CONTROL

1. Michael Veale, 'Privacy is not the problem with the Apple–Google contact-tracing app', *The Guardian*, 1 July 2020. Available at: https://www.theguardian.com/commentisfree/2020/jul/01/apple-google-contact-tracing-app-tech-giant-digital-rights. Accessed 1 July 2020.
2. Jim Dunton, '"Dom is full of amazing ideas": MOD will have to back Cummings reform plan says defence secretary', Civil Service World, 20 December 2019. Available at: https://www.civilserviceworld.com/articles/news/%E2%80%98dom-full-amazing-ideas%E2%80%99-%E2%80%93-mod-will-have-back-cummings-reform-plan-says-defence. Accessed 2 January 2020.
3. 'The modern workplace: San Francisco jobs, skills, and opportunities in the age of AI', Accenture and San Francisco Chamber of Commerce,

May 2019. Available at: https://www.accenture.com/us-en/insights/future-workforce/sf-tech-jobs. Accessed 20 December 2019.

4. Ryan Nunn and Jay Shambaugh, 'San Francisco: where a six-figure salary is "low income"', BBC News, 10 July 2018. Available at: https://www.bbc.co.uk/news/world-us-canada-44725026. Accessed 20 December 2019.
5. See #vetswhocode. Available at: https://vetswhocode.io. Accessed 22 May 2019.
6. 'Tech that connects us makes us better humans', *Wired*, 14 May 2019. Available at: https://www.wired.com/story/why-we-love-tech-better-humans. Accessed 22 May 2019.
7. 'About Operation War Diary'. Available at: https://www.operationwardiary.org/#/about. Accessed 22 May 2019.
8. Daisuke Wakabayashi and Scott Shane, 'Google will not renew Pentagon contract that upset employees', *New York Times*, 1 June 2018. Available at: https://www.nytimes.com/2018/06/01/technology/google-pentagon-project-maven.html. Accessed 22 December 2019.
9. Peter Thiel, 'Opinion: good for Google, bad for America', *New York Times*, 1 August 2019. Available at: https://www.nytimes.com/2019/08/01/opinion/peter-thiel-google.html. Accessed 9 January 2020.
10. Steve Blank, 'Why the government isn't a bigger version of a startup', War on the Rocks, 11 November 2019. Available at: https://warontherocks.com/2019/11/why-the-government-isnt-a-bigger-version-of-a-startup. Accessed 28 December 2019.
11. Elsa Kania and Emma Moore, 'Great power rivalry is also a war for talent', Defense One, 19 May 2019. Available at: https://www.defenseone.com/ideas/2019/05/great-power-rivalry-also-war-talent/157103/?oref=d-river. Accessed 22 May 2019.
12. Facebook now have products that 'empower more than 2 billion people around the world to share ideas'. See Facebook corporate website: https://about.fb.com/company-info. Accessed 23 December 2019.
13. ICT data for the world, by geographic regions, by urban/rural area and by level of development for the following indicators (2005–20; Excel). Available at: https://www.itu.int/en/ITU-D/Statistics/Documents/facts/ITU_regional_global_Key_ICT_indicator_aggregates_Nov_2020.xlsx. The ICT's main statistics page is available here: https://www.itu.int/en/ITU-D/Statistics/Pages/stat/default.aspx. Both pages accessed on 26 November 2021.
14. See, for example, these two articles that describe the impact of smart city technology in Canada and China: Ava Kofman, 'Google's

"smart city of surveillance" faces new resistance in Toronto', The Intercept, 13 November 2018. Available at: https://theintercept.com/2018/11/13/google-quayside-toronto-smart-city; and Isobel Cockerell, 'Inside China's massive surveillance operation', *Wired*, 9 May 2019. Available at: https://www.wired.com/story/inside-chinas-massive-surveillance-operation. Both articles accessed 2 January 2020.

15. Tom Simonite, 'Artificial Intelligence is coming for our faces', *Wired*, 24 June 2019. Available at: https://www.wired.com/story/artificial-intelligence-fake-fakes. Accessed 23 December 2019.
16. Sarah Frier, 'Samsung phone users perturbed to find they can't delete Facebook', Bloomberg, 8 January 2019. Available at: https://www.bloomberg.com/news/articles/2019-01-08/samsung-phone-users-get-a-shock-they-can-t-delete-facebook. Accessed 2 January 2020.
17. The Philippines, for example, has become a huge market for smartphone providers who link up with local network providers to offer cheap phones with pre-configured information architectures pre-loaded on to them. For market share by vendor, see 'The Philippines is now the fastest growing smartphone market in ASEAN', Marketing, 24 June 2016. Available at: https://www.marketing-interactive.com/philippines-now-fastest-growing-smartphone-market-asean. Accessed 28 December 2019.
18. Jonathan Ong and Jason Cabañes, 'Architects of networked disinformation: behind the scenes of troll accounts and fake news production in the Philippines', 5 February 2018. Available at: http://newtontechfordev.com/newton-tech4dev-research-identifies-ad-pr-executives-chief-architects-fake-news-production-social-media-trolling. Accessed 2 January 2020.
19. More detail on the information environment in Syria can be found in: Freedom House, 'Freedom on the net 2018: Syria', 1 November 2018. Available at: https://www.refworld.org/docid/5be16af6116.html. Accessed 16 July 2019.
20. Megha Rajagopalan, 'The histories of today's war are being written on Facebook and YouTube: but what happens when they get taken down?', Buzzfeed News, 22 December 2018. Available at: https://www.buzzfeednews.com/article/meghara/facebook-youtube-icc-war-crimes. Accessed 17 July 2019.
21. Casey Newton, 'Facebook is losing control of the narrative – and maybe the platform', The Verge, 20 March 2018. Available at: https://www.theverge.com/2018/3/20/17140490/facebook-cambridge-analytica-data-crisis. Accessed 7 January 2020.

22. Andy Greenberg, 'How an entire nation became Russia's test lab for Cyberwar', *Wired*, 20 June 2017. Available at: https://www.wired.com/story/russian-hackers-attack-ukraine. Accessed 3 January 2020.
23. Dan Sabbagh, 'Cambridge Analytica parent company had access to secret MOD information', *The Guardian*, 29 March 2018. Available at: https://www.theguardian.com/uk-news/2018/mar/29/cambridge-analytica-predecessor-had-access-to-secret-mod-information. Accessed 6 January 2020.
24. 'Section 7, influence in foreign relations', Report of the Culture, Media and Sport Select Committee, House of Commons, 18 February 2019. Available at: https://publications.parliament.uk/pa/cm201719/cmselect/cmcumeds/1791/179102.htm. Accessed 6 January 2020.
25. Michal Kosinski and Renaud Lambiotte, 'Tracking the digital footprints of personality', *Proceedings of the IEEE* 102:12 (December 2014). Available at: https://www.gsb.stanford.edu/faculty-research/publications/tracking-digital-footprints-personality. Accessed 7 January 2020.
26. 'Report of the Culture, Media and Sport Select Committee', House of Commons, 18 February 2019. Available at: https://publications.parliament.uk/pa/cm201719/cmselect/cmcumeds/1791/179107.htm#_idTextAnchor034. Accessed 6 January 2020.
27. 'Interim report of the Culture, Media and Sport Select Committee', House of Commons, 24 July 2018, p. 31. Available at: https://publications.parliament.uk/pa/cm201719/cmselect/cmcumeds/363/363.pdf. Accessed 7 January 2020.
28. Watch the interview, 'Propaganda machine: the military roots of Cambridge Analytica's psychological manipulation of voters', Democracy Now, 7 January 2020. Available at: https://www.democracynow.org/2020/1/7/cambridge_analytica_data_manipulation_john_bolton. Accessed 7 January 2020.
29. Craig Silverman, Jane Lytvynenko and William Kung, 'Disinformation for hire: how a new breed of PR firms is selling lies online', Buzzfeed, 6 January 2020. Available at: https://www.buzzfeednews.com/article/craigsilverman/disinformation-for-hire-black-pr-firms. Accessed 8 January 2020.
30. Influencer marketing platform market size worldwide 2017–27, Statistica, 19 October 2021. Available at: https://www.statista.com/statistics/1036560/global-influencer-marketing-platform-market-size. Accessed 26 November.

31. Digital advertising spending worldwide from 2019 to 2024, Statistica, 28 May 2021. Available at: https://www.statista.com/statistics/237974/online-advertising-spending-worldwide. Accessed 26 November 2021.
32. Jesse Frederik and Maurits Martijn, 'The new dot com bubble is here: it's called online advertising', The Correspondent, 6 November 2019. Available at: https://thecorrespondent.com/100/the-new-dot-com-bubble-is-here-its-called-online-advertising/13228924500-22d5fd24. Accessed 7 January 2020.
33. Ibid.
34. Sharon Weinberger, 'Meet America's newest military giant: Amazon', *MIT Technology Review*, 8 October 2019. Available at: https://www.technologyreview.com/s/614487/meet-americas-newest-military-giant-amazon. Accessed 22 November 2019.
35. Glenn Greenwald, 'Revealed: how US and UK spy agencies defeat internet privacy and security', *The Guardian*, 6 September 2013. Available at: https://www.theguardian.com/world/2013/sep/05/nsa-gchq-encryption-codes-security. Accessed 9 January 2020.
36. Glenn Greenwald, 'How the NSA tampers with US-made internet routers', *The Guardian*, 12 May 2014. Available at: https://www.theguardian.com/books/2014/may/12/glenn-greenwald-nsa-tampers-us-internet-routers-snowden. Accessed 9 January 2020.
37. Alex Purcell et al., 'The Snowden files: facts and figures – video animation', *The Guardian*, 2 December 2013. Available at: https://www.theguardian.com/world/video/2013/dec/02/the-snowden-files-guide-surveillance-industry-video. Accessed 9 January 2020.
38. China's approach being particularly coercive; see Chris Buckley and Paul Mozur, 'How China uses high-tech surveillance to subdue minorities', *New York Times*, 22 May 2019. Available at: https://www.nytimes.com/2019/05/22/world/asia/china-surveillance-xinjiang.html. Accessed 9 January 2020.
39. Doug Irving, 'Four ways 3D printing may threaten security', RAND Corporation, 8 May 2018. Available at: https://www.rand.org/blog/articles/2018/05/four-ways-3d-printing-may-threaten-security.html. Accessed 15 January 2019.
40. 'Russia's controversial "Sovereign Internet" law comes into force', Radio Free Europe/Radio Liberty, 1 November 2019. Available at: https://www.rferl.org/a/russia-s-controversial-sovereign-internet-law-comes-into-force/30247754.html. Accessed 8 January 2020.
41. 'Russia bans sale of smartphones without Russian apps', Moscow Times, 2 December 2019. Available at: https://www.

themoscowtimes.com/2019/12/02/russia-bans-iphone-sale-without-russian-apps-a68313. Accessed 8 January 2020.

42. 'Russia disconnects from internet in tests as it bolsters security – RBC daily', Reuters, 22 July 2021. Available at: https://www.reuters.com/technology/russia-disconnected-global-internet-tests-rbc-daily-2021-07-22. Accessed 23 July 2021.
43. Alison Mutler, 'Pro-Orbán media moguls who destroyed Hungary's media now targeting European outlets', Coda, 28 June 2019. Available at: https://codastory.com/disinformation/orban-media-moguls-targeting-european-outlets. Accessed on 8 January 2020.
44. Umesh Moramudali, 'Is Sri Lanka really a victim of China's "debt trap"?', The Diplomat, 14 May 2019. Available at: https://thediplomat.com/2019/05/is-sri-lanka-really-a-victim-of-chinas-debt-trap. Accessed 8 January 2020.
45. Trym Aleksander Eiterjord, 'Checking in on China's nuclear icebreaker', The Diplomat, 5 September 2019. Available at: https://thediplomat.com/2019/09/checking-in-on-chinas-nuclear-icebreaker. Accessed 21 July 2020.
46. David Auerswald, 'China's multifaceted Arctic strategy', War on the Rocks, 24 May 2019. Available at: https://warontherocks.com/2019/05/chinas-multifaceted-arctic-strategy. Accessed 21 July 2020.
47. 'Australia investigates alleged Chinese plot to install spy MP', BBC News, 25 November 2019. Available at: https://www.bbc.co.uk/news/world-australia-50541082. Accessed 21 July 2020.
48. Joan Tilouine and Ghalia Kadiri, 'A Addis-Abeba, le siège de l'Union africaine espionné par Pékin', *Le Monde*, 27 January 2018. Available at: https://www.lemonde.fr/afrique/article/2018/01/26/a-addis-abeba-le-siege-de-l-union-africaine-espionne-par-les-chinois_5247521_3212.html. Accessed 8 January 2020.
49. 'The 5G economy: how 5G will impact global industries, the economy, and you', *MIT Technology Review*, 1 March 2017. Available at: https://www.technologyreview.com/s/603770/the-5g-economy-how-5g-will-impact-global-industries-the-economy-and-you. Accessed 8 January 2020.
50. Lara Seligman, 'America's answer to Huawei', *Foreign Policy*, 1 October 2019. Available at: https://foreignpolicy.com/2019/10/01/america-answer-to-huawei-pentagon-carriers-5g-race-china. Accessed 8 January 2020.
51. 'China and the United States race for 5G', Oxford Analytica Daily Brief, 28 February 2018.

52. 'Men Against Fire', *Black Mirror*, Series 3, Episode 3, first broadcast on Netflix.com. Now available at: https://www.youtube.com/watch?v=2KP_ipE5mLY. Accessed 9 January 2020.
53. Interview with Thomas Reardon, CEO of CTRL-Labs, 'Neural interface technology and the future of human–computer interaction', Hidden Forces Podcast, 11 March 2019. Available at: https://www.youtube.com/watch?v=cAbGHvf9Ndw. Accessed 9 January 2020.
54. John Amble, 'How science can optimize cognitive performance on the battlefield', Modern War Institute Podcast, 13 June 2019. Available at: https://mwi.usma.edu/mwi-podcast-science-can-optimize-cognitive-performance-battlefield. Accessed 10 January 2020.
55. Todd South, 'Soldiers, marines try out new device that puts "mixed reality", multiple functions into warfighter's hands', *Army Times*, 8 April 2019. Available at: https://www.armytimes.com/news/your-army/2019/04/08/soldiers-marines-try-out-new-device-that-puts-mixed-reality-multiple-functions-into-warfighters-hands. Accessed 10 January 2020.
56. Emily Singer, 'The maestro of memory manipulation', Quanta Magazine, 23 June 2016. Available at: https://www.quantamagazine.org/the-maestro-of-memory-manipulation-20160623. Accessed 10 January 2020.
57. Henry Mance, 'Britain has had enough of experts, says Gove', *Financial Times*, 3 June 2016. Available at: https://www.ft.com/content/3be49734-29cb-11e6-83e4-abc22d5d108c. Accessed 13 January 2020.
58. Jennifer Lai, 'Information wants to be free and expensive', Forbes.com, 20 July 2019. Available at: http://fortune.com/2009/07/20/information-wants-to-be-free-and-expensive. Accessed 13 May 2019.
59. Richard Barbrook and Andy Cameron, 'The Californian ideology', *Mute* 1(3) (1 September 1995). Available at: https://www.metamute.org/editorial/articles/californian-ideology. Accessed 11 January 2020.

CONCLUSION

1. Shane Harris and Yochi Dreazen, 'U.S. intel sources: Russian invasion of eastern Ukraine increasingly likely', *Foreign Policy*, 27 March 2014. Available at: https://foreignpolicy.com/2014/03/27/u-s-intel-sources-russian-invasion-of-eastern-ukraine-increasingly-likely. Accessed 31 October 2021.
2. In 1994, Ukraine agreed to give up the nuclear weapons it had inherited following the collapse of the Soviet Union on the basis that

its sovereignty would be guaranteed by the Russian, American and British governments.

3. Halya Coynash, 'New United Russia party MP confirms that the fighting in the Donbas is by "Russian forces"', Kharkiv Human Rights Protection Group, 29 October 2021. Available at: https://khpg.org/en/1608809673. Accessed 31 October 2021.
4. 'DISINFO: the EU triggered conflict in Ukraine in 2014', EU vs DiSiNFO 151, 21 May 2019. Available at: https://euvsdisinfo.eu/report/the-eu-triggered-conflict-in-ukraine-in-2014. Accessed 31 October 2021.
5. Charles Grant, 'Is the EU to blame for the crisis in Ukraine?', Centre for European Reform, 1 June 2016. Available at: https://www.cer.eu/insights/eu-blame-crisis-ukraine. Accessed 31 October 2021.

EPILOGUE

1. '"We never got it. Not even close": Afghanistan veterans reflect on 20 years of war', Politico Magazine, 10 September 2021. Available at: https://www.politico.com/news/magazine/2021/09/10/politico-mag-afghan-vets-roundtable-506989. Accessed 30 October 2021.
2. Stuart Scheller, 'To the American leadership. Very respectfully, US', Facebook, 26 August 2021. Available at: https://www.facebook.com/stuart.scheller/videos/561114034931173/?t=238. Accessed 30 October 2021.
3. Stuart Scheller, 'Your move', YouTube, 29 August 2021. Available at: https://www.youtube.com/watch?v=lR7jBsR0D10&t=495s. Accessed 30 October 2021.
4. Jeff Schogol, 'Leaked documents reveal just how concerned the Marine Corps was about Lt. Col. Stuart Scheller's call for "revolution"', Task and Purpose, 17 October 2021. Available at: https://taskandpurpose.com/news/marine-corps-lt-col-stuart-scheller-court-martial. Accessed 30 October 2021.
5. Mike Wendling, 'QAnon: what is it and where did it come from?', BBC News, 6 January 2021. Available at: https://www.bbc.co.uk/news/53498434. Accessed 30 October 2021.
6. The video was posted on Twitter by Katherine Denkinson at 20:23 on 9 August 2021. Available at: https://twitter.com/KDenkWrites/status/1424813677849415685?s=20. Accessed 30 October 2021.
7. Ibid.

APPENDIX

1. Jason Bloomberg, 'Digitization, digitalization, and digital transformation: confuse them at your peril', *Forbes*, 29 April 2018. Available at: https://www.forbes.com/sites/jasonbloomberg/2018/04/29/digitization-digitalization-and-digital-transformation-confuse-them-at-your-peril/?sh=67d95fe32f2c. Accessed 26 November 2021.
2. Digitalisation, Gartner Glossary. Available at: https://www.gartner.com/en/information-technology/glossary/digitalization. Accessed 26 November 2021.
3. Jeremy W. Peters, 'Bannon's worldview: dissecting the message of "The Fourth Turning"', *New York Times*, 8 April 2017. Available at: https://www.nytimes.com/2017/04/08/us/politics/bannon-fourth-turning.html. Accessed 8 June 2020.
4. Matt Burgess, 'What is the internet of things? WIRED explains', *Wired*, 16 February 2018. Available at: https://www.wired.co.uk/article/internet-of-things-what-is-explained-iot. Accessed 10 July 2020.
5. Ryan Singel, 'Are you ready for Web 2.0?', *Wired*, 6 October 2005. Available at: https://www.wired.com/2005/10/are-you-ready-for-web-2-0. Accessed 10 July 2020.

BIBLIOGRAPHY

Agostinho, D., S. Gade, N. B. Thylstrup and K. Vee (eds) (2021). *(W)archives: Archival Imaginaries, War, and Contemporary Art*. Berlin: Sternberg Press.

Agre, P. E. (1994). 'Understanding the digital individual'. *The Information Society* 10(2): 73–6.

Alberts, D., et al. (eds) (2001). *Understanding Information Age Warfare*. Washington, DC: CCRP Publication Series.

Allison, W. T. (2019). 'Provisional healing: Vietnam memorials and the limits of memory'. In G. W. Jensen and M. M. Stith (eds), *Beyond the Quagmire: New Interpretations of the Vietnam War*. Denton, TX: University of North Texas Press. pp. 359–90.

Almohammad, A. and C. Winter (2019). 'From Directorate of Intelligence to directorate of everything: the Islamic State's emergent Amni–media nexus'. *Perspectives on Terrorism* 13(1): 41–53.

Amoore, L. (2009). 'Algorithmic war: everyday geographies of the War on Terror'. *Antipode* 41(1): 49-69.

Andén-Papadopoulos, K. (2009). 'US soldiers imaging the Iraq War on YouTube'. *Popular Communication* 7(1): 17–27.

Andrejevic, M. (2014). 'The big data divide'. *International Journal of Communication* 8: 1673–89.

Applebaum, A. (2020). *Twilight of Democracy: The Seductive Lure of Authoritarianism*. New York: Doubleday.

Arlen, M. J. (1966). *Living-Room War*. New York: Viking Press.

Arquilla, J. (2021). *Bitskrieg: The New Challenge of Cyberwarfare*. Cambridge: Polity Press.

Arquilla, J. and D. A. Borer (eds) (2007). *Information Strategy and Warfare: A Guide to Theory and Practice*. London: Routledge.

Arquilla, J. and D. Ronfeldt (1993). 'Cyberwar is coming'. *Comparative Strategy* 12(2): 141–65.

Aupers, S. (2012). '"Trust no one": modernization, paranoia and conspiracy culture'. *European Journal of Communication* 27(1): 22–34.

Avant, D. (2005). *The Market for Force: The Consequences of Privatizing Security*. Cambridge: Cambridge University Press.

Ball, J. (2020). *The System: Who Owns the Internet, and How It Owns Us*. London: Bloomsbury Publishing.

Bartlett, F. C. (1932). *Remembering: A Study in Experimental and Social Psychology*. Cambridge: Cambridge University Press.

Baudrillard, J. (1991/5). *The Gulf War Did Not Take Place*. Translated by Paul Patton. Sydney: Power Publications

—— (1994). *The Illusion of the End*. Cambridge: Polity Press.

Bell, M. (2008). 'The death of news'. *Media, War & Conflict* 1(2): 221–31.

Bennett, H. (2011). 'Soldiers in the court room: the British Army's part in the Kenya Emergency under the legal spotlight'. *The Journal of Imperial and Commonwealth History* 39(5): 717–30.

—— (2012). *Fighting the Mau Mau: The British Army and Counter-Insurgency in the Kenya Emergency*. Cambridge: Cambridge University Press.

Berman, E., J. H. Felter and J. S. Shapiro (2018). *Small Wars, Big Data: The Information Revolution in Modern Conflict*. Princeton: Princeton University Press.

Blanken, L. (2012). *Rational Empires: Institutional Incentives and Imperial Expansion*. Chicago: Chicago University Press.

Blight, D. (2012). 'From the Civil War to civil rights and beyond: how Americans have remembered their deepest conflict'. Paper presented at The Future of Memory Conference, University of Konstanz, 5 July 2012.

Bolt, N. (2012). *The Violent Image: Insurgent Propaganda and the New Revolutionaries*. London: Hurst & Co.

Bousquet, A. (2008). 'Chaoplexic warfare'. *International Affairs* 84(5): 915–29.

—— (2018). *The Eye of War: Military Perception from the Telescope to the Drone*. Minneapolis: University of Minneapolis Press.

Bowker, G. C. (2007). 'The past and the internet'. In J. Karaganis (ed.), *Structures of Participation in Digital Culture*. New York: Social Science Research Council, pp. 20–36.

—— (2016). 'Just what are we archiving?' *Limn*, Issue 6, 'The Total Archive'. https://limn.it/articles/just-what-are-we-archiving.

Bowker, G. C., K. Baker, F. Millerand and D. Ribes (2010). 'Toward information infrastructure studies: ways of knowing in a networked environment'. In J. Hunsinger, L. Kalstrup and M. Allen (eds), *International Handbook of Internet Research*. Heidelberg: Springer, pp. 97–117.

Bowker, G. C. and S. L. Star (2000). *Sorting Things Out: Classification and Its Consequences*. Cambridge, MA: MIT Press.

Braestrup, P. (1983). *Big Story: How the American Press and Television Reported and Interpreted the Crisis of Tet 1968 in Vietnam and Washington*. Abridged edn. New Haven: Yale University Press.

Brandtzaeg, P. B. and M. Lüders (2018). 'Time collapse in social media: extending the context collapse'. *Social Media + Society* 4(1).

Briant, E. L. (2015a). *Propaganda and Counter-Terrorism*. Manchester: Manchester University Press.

—— (2015b). 'Allies and audiences: evolving strategies in defense and intelligence propaganda'. *The International Journal of Press/Politics* 20(2): 145–65.

—— (2018). 'Pentagon ju-jitsu: reshaping the field of propaganda'. *Critical Sociology* 45(3): 361–78.

—— (2019). 'LeaveEU: dark money, dark ads and data crimes'. In P. Baines, N. Snow and N. O'Shaughnessy (eds), *SAGE Handbook of Propaganda*. London: SAGE. pp. 532–49.

Bratton B. H. (2016). *The Stack: On Software and Sovereignty*. Cambridge, MA: MIT Press.

Brauman, R. (2019). *Humanitarian Wars? Lies and Brainwashing*. Translated by Nina Friedman. London: C. Hurst & Co.

Bridle, A. (2019). *New Dark Age: Technology and the End of the Future*. London: Verso.

Broadbent, S. and C. Lobet-Maris (2014). 'Towards a grey ecology'. In L. Floridi (ed.), *The Online Manifesto*. Cham: Springer, pp. 111–24.

Brooks, R. (2016). *How Everything Became War and the Military Became Everything*. New York: Simon & Schuster.

Brose, C. (2020). *The Kill Chain: Defending America in the Future of High-Tech Warfare*. New York: Hachette Books.

Brown, S. and A. Hoskins (2010). 'Terrorism in the new memory ecology: mediating and remembering the 2005 London bombings'. *Behavioral Sciences of Terrorism and Political Aggression* 2(2): 87–107.

Carr, A. (2018). 'It's about time: strategy and temporal phenomena'. *Journal of Strategic Studies* 44(3): 303–24.

Catignani, S. (2012). '"Getting COIN" at the tactical level in Afghanistan: reassessing counter-insurgency adaptation in the British Army'. *Journal of Strategic Studies* 35(4): 513–39.

Chamayou, G. (2012). *Manhunts: A Philosophical History*. Princeton: Princeton University Press.

—— (2015). *Drone Theory*. London: Penguin.

Chaudhuri, R. and T. Farrell (2011). 'Campaign disconnect: operational progress and strategic obstacles in Afghanistan, 2009–2011'. *International Affairs* 87(2): 271–96.

Cheney-Lippold, J. (2017). *We Are Data: Algorithms and the Making of Our Digital Selves*. New York: NYU Press.

Chéroux, C. (2012). 'The déjà vu of September 11: an essay on intericonicity'. Translated by Hillary Goidell. In F. Hoffman (ed.), *The Uncanny Familiar: Images of Terror*. Cologne: Walther König, bilingual edition, pp. 261–87.

Chotikul, D. (1986). 'The Soviet theory of reflexive control in historical and psychological perspective: a preliminary study'. Monterey, Naval Postgraduate School.

Cohen, E. (2018). *The Political Value of Time: Citizenship, Duration, and Democratic Justice*. Cambridge: Cambridge University Press.

Coker, C. (2012). *Warrior Geeks: How 21st Century Technology is Changing the Way We Fight and Think about War*. London: Hurst & Co.

Connerton, P. (2008). 'Seven types of forgetting'. *Memory Studies* 1(1): 59–70.

Cormac, R. and R. J. Aldrich (2018). 'Grey is the new black: covert action and implausible deniability'. *International Affairs* 94(3): 477–94.

Cornish, P. and A. Dorman (2009). 'Blair's wars and Brown's budgets: from Strategic Defence Review to strategic decay in less than a decade'. *International Affairs* 85(2): 247–61.

Cottle, S. (2006). *Mediatized Conflict: Developments in Media and Conflict Studies*. Maidenhead: Open University Press.

Couldry, N. and A. Hepp (2017). *The Mediated Construction of Reality*. Cambridge: Polity Press.

Cronin, A. K. (2020). *Power to the People: How Open Technological Innovation Is Arming Tomorrow's Terrorists*. New York: Oxford University Press.

Davies, W. (2018). *Nervous States: How Feeling Took Over the World*. London: Jonathan Cape.

De Franco, C. (2012). *Media Power and the Transformation of War*. London: Palgrave Macmillan.

Dear, K. (2019). 'Will Russia rule the world through AI?' *The RUSI Journal* 164(5–6): 36–60.

Der Derian, J. (2009). *Virtuous War: Mapping the Military–Industrial–Media–Entertainment Network*. London: Routledge.

Devji, F. (2005). *Landscapes of the Jihad: Militancy, Morality, Modernity*. London: Hurst & Co.

—— (2009). *The Terrorist in Search of Humanity: Militant Islam and Global Politics*. London: Hurst & Co.

Dixon, P. (2019). 'Frock coats against brass hats? Politicians, the military and the war in Afghanistan 2001–2014'. *Parliamentary Affairs* 73(3): 651–91.

Dobbins, C., M. Merabti, P. Fergus and D. Llewellyn-Jones (2013). 'Creating human digital memories with the aid of pervasive mobile devices'. *Pervasive and Mobile Computing* 12: 160–78. DOI: http://dx.doi.org/10.1016/j.pmcj.2013.10.009

Domby, A. (2020). *The False Cause: Fraud, Fabrication and White Supremacy in Confederate Memory*. Charlottesville: University of Virginia Press.

Echevarria, A. J. (2016). *Operating in the Gray Zone: An Alternative Paradigm for U.S. Military Strategy*. Washington, DC: Strategic Studies Institute.

Edgerton, D. (2008). *Shock of the Old: Technology and Global History since 1900*. London: Profile Books.

Edmunds, T. (2012). 'British civil–military relations and the problem of risk'. *International Affairs* 88(2): 265–82.

Eriksson Krutrök, M. and S. Lindgren (2018). 'Continued contexts of terror: analyzing temporal patterns of hashtag co-occurrence as discursive articulations'. *Social Media + Society* 4(4).

Ernst, W. (2004). 'The archive as metaphor'. *Open* 7: 46–53.

Eyal, G. (2019). *The Crisis of Expertise*. Cambridge: Polity.

Farkas, J. and J. Schou (2018). 'Fake news as a floating signifier: hegemony, antagonism and the politics of falsehood'. *Javnost: The Public* 25(3): 298–314.

Farrell, T. (2017). *Unwinnable: Britain's War in Afghanistan, 2001–2014*. London: Bodley Head.

Feldman, A. (2009). 'The structuring enemy and archival war'. *PMLA* 124(5): 1704–13.

FitzGerald, B. and J. Parziale (2017). 'As technology goes democratic, nations lose military control'. *Bulletin of the Atomic Scientists* 73(2): 102–7.

Floridi, L. (2013). 'Hyperhistory and the philosophy of information policies'. EU Onlife Initiative. https://ec.europa.eu/digital-agenda/sites/digital-agenda/files/Onlife_Initiative.pdf

Foer, F. (2017). *World Without Mind: The Existential Threat of Big Tech*. London: Jonathan Cape.

Ford, M. (2012). 'Finding the target, fixing the method: methodological tensions in insurgent identification'. *Studies in Conflict & Terrorism* 35(2): 113–34.

—— (2019). 'The epistemology of lethality: bullets, knowledge trajectories, kinetic effects'. *European Journal of International Security* 5(1): 77–93.

—— (2021). 'Review essay of *Changing of the Guard* and *Blood, Metal and Dust*'. *Journal of Strategic Studies*: 1–11.

Ford, M. and J. Michaels (2011). 'Bandwagonistas: rhetorical redescription, strategic choice and the politics of counterinsurgency'. *Small Wars & Insurgencies* 22(2): 352–84.

Freedman, L. and J. Michaels (eds) (2013). *Scripting Middle East Leaders: The Impact of Leadership Perceptions on US and UK Foreign Policy*. London: Bloomsbury.

Freedman, L. (2017). *The Future of War: A History*. London: Penguin.

Fridman, O. (2018). *Russian Hybrid Warfare: Resurgence and Politicisation*. London: Hurst & Co.

Fukuyama, F. (1992). *The End of History and the Last Man*. New York: Free Press.

Fuller, M. (2007). *Media Ecologies*. London: MIT Press.

Fuller, M. and A. Goffey (2012). *Evil Media*. Cambridge, MA: MIT Press.

Galeotti, M. (2016). 'Hybrid, ambiguous, and non-linear? How new is Russia's "new way of war"?' *Small Wars & Insurgencies* 27(2): 282–301.

—— (2019). *Russian Political War: Moving beyond the Hybrid*. London: Routledge.

Ganguly, D. (2016). *This Thing Called the World: The Contemporary Novel as Global Form*. Durham, NC: Duke University Press.

Gilroy, P. (2006). 'Multiculture in times of war: an inaugural lecture given at the London School of Economics'. *Critical Quarterly* 48(4): 27–45.

Goldhaber, M. H. (1997). 'The attention economy and the net'. *First Monday* 2(4). DOI: https://doi.org/10.5210/fm.v2i4.519

Gregory, D. (2011). 'From a view to a kill: drones and late modern war'. *Theory, Culture and Society* 28(7–8): 188–215.

Griffiths, J. (2019). *The Great Firewall of China: How to Build and Control an Alternative Version of the Internet*. London: Zed Books.

Grusin, R. (2015). 'Radical mediation'. *Critical Inquiry* 42(1): 124–48.

Guriev, S. and D. Treisman (2019). 'Informational autocrats'. *Journal of Economic Perspectives* 33(4): 100–27.

Hallin, D. C. (1986). *The Uncensored War: The Media and Vietnam*. New York: Oxford University Press.

Hammes, T. X. (2004). *The Sling and the Stone: On War in the 21st Century*. Minneapolis: Zenith Press.

Happer, C. and A. Hoskins (2022). 'Hacking the archive: media, memory and history in the post-trust era'. In M. Moss and D. Thomas (eds), *Post-Truth in the Archives*. Oxford: Oxford University Press.

Harari, Y. N. (2008). *The Ultimate Experience: Battlefield Revelations and the Making of Modern War Culture, 1450–2000*. Basingstoke: Palgrave Macmillan.

Harkness, T. (2017). *Big Data: Does Size Matter?* London: Bloomsbury Sigma.

Hashim, A. S. (2018). *The Caliphate at War: The Ideological, Organisational and Military Innovations of Islamic State*. London: Hurst & Co.

Hauter, J. (2021). 'Forensic conflict studies: making sense of war in the social media age'. *Media, War & Conflict*. https://journals.sagepub.com/doi/10.1177/17506352211037325

Head, H. (1920). *Studies in Neurology*. New York: Oxford University Press.

Hepp, A. (2019). *Deep Mediatization*. London, Routledge.

Hilberg, R. (1993). *Perpetrators, Victims, Bystanders: The Jewish Catastrophe, 1933–1945*. New York: Harper Perennial.

Hoffman, F. (2007). *Conflict in the 21st Century: The Rise of Hybrid Wars*. Arlington, VA: Potomac Institute for Policy Studies.

Hoffmann, A. L., N. Proferes and M. Zimmer (2016). '"Making the world more open and connected": Mark Zuckerberg and the discursive construction of Facebook and its users'. *New Media & Society* 20(1): 199–218.

Horowitz, M. C. (2019a). 'When speed kills: lethal autonomous weapon systems, deterrence and stability'. *Journal of Strategic Studies* 42(6): 764–88.

—— (2019b). 'Artificial intelligence and nuclear stability'. In V. Boulanin (ed.), *The Impact of Artificial Intelligence on Strategic Stability and Nuclear Risk*. Stockholm, SIPRI, pp. 79–90.

Hoskins, A. (2004). *Televising War: From Vietnam to Iraq*. London: Continuum.

—— (2011a). '7/7 and connective memory: interactional trajectories of remembering in post-scarcity culture'. *Memory Studies* 4(3): 269–80.

—— (2011b). 'Media, memory, metaphor: remembering and the connective turn'. *Parallax* 17(4): 19–31.

—— (2015). 'Archive me! Media, memory, uncertainty'. In A. Hajek, C. Lohmeier and C. Pentzold (eds), *Memory in a Mediated World: Remembrance and Reconstruction*. Basingstoke: Palgrave Macmillan, pp. 13–35.

—— (2018). 'The restless past: an introduction to digital memory and media'. In A. Hoskins (ed.), *Digital Memory Studies: Media Pasts in Transition*. New York: Routledge, pp. 1–24.

Hoskins, A. and M. Ford (2017). 'Flawed, yet authoritative? Organisational memory and the future of official military history after Chilcot'. *British Journal for Military History* 3(2): 119–32.

Hoskins, A. and A. Holdsworth (2015). 'Media archaeology of/in the museum'. In M. Henning (ed.), *Museum Media*. Oxford: Wiley-Blackwell, pp. 23–42.

Hoskins, A. and W. Merrin (2021). 'Remember Afghanistan?' *Journal of Digital War*.

Hoskins, A. and B. O'Loughlin (2009). *Television and Terror: Conflicting Times and the Crisis of News Discourse*. Basingstoke: Palgrave Macmillan.

—— (2010). *War and Media: The Emergence of Diffused War*. Cambridge: Cambridge University Press.

—— (2015). 'Arrested war: the third phase of mediatization'. *Information, Communication & Society* 18(11): 1320–38.

Hoskins, A. and P. Shchelin (2018). 'Information war in the Russian media ecology: the case of the Panama Papers'. *Continuum* 32(1): 1–17.

Hoskins, A. and J. Tulloch (2016). *Risk and Hyperconnectivity: Media and Memories of Neoliberalism*. Oxford: Oxford University Press.

Howard, M. (2001). *War in European History*. Oxford: Oxford University Press.

Hughes, G. (2012). *My Enemy's Enemy: Proxy Warfare in International Politics*. Brighton: Sussex Academic Press.

Hundley, R. (1999). *Past Revolutions, Future Transformations: What Can the History of Revolutions in Military Affairs Tell Us about Transforming the US Military?* Santa Monica, CA: RAND.

Huntington, S. P. (1957). *The Soldier and the State: The Theory and Politics of Civil–Military Relations*. Cambridge, MA: Harvard University Press.

—— (1996). *The Clash of Civilizations and the Remaking of World Order*. New York: Simon & Schuster.

Huyssen, A. (2000). 'Present pasts: media, politics, amnesia'. *Public Culture* 12(1): 21–38.

—— (2003). *Present Pasts: Urban Palimpsests and the Politics of Memory*. Stanford: Stanford University Press.

Ibrahim, Y. (2009). 'City under siege: narrating Mumbai through nonstop capture'. *Culture Unbound: Journal of Current Cultural Research* 1(15): 385–99.

Ingram, H., C. Whiteside and C. Winter (2020). *The ISIS Reader: Milestone Texts of the Islamic State Movement*. London: Hurst & Co.

Inkster, N. (2020). *The Great Decoupling: China, America and the Struggle for Technological Supremacy*. London: Hurst & Co.

Innes, M. (2021). *Streets Without Joy – a political history of sanctuary and war 1959-2009*. London, Hurst & Co.

Jackson, V. (2017). 'Tactics of strategic competition: gray zones, redlines, and conflicts before war'. *Naval War College Review* 70(3): 39–61.

Jacobsen, A. (2021). *First Platoon: A Story of Modern War in the Age of Identity Dominance*. New York: Dutton.

Jankowicz, N. (2020). *How to Lose the Information War: Russia, Fake News and the Future of Conflict*. London: I. B. Tauris.

Johnson, J. (2021). *Artificial Intelligence and the Future of Warfare: The USA, China, and Strategic Stability*. Manchester: Manchester University Press.

Jonsson, O. and R. Seely (2015). 'Russian full-spectrum conflict: an appraisal after Ukraine'. *The Journal of Slavic Military Studies* 28(1): 1–22.

Kagan, F. W. (2006). *Finding the Target: The Transformation of American Military Power*. New York: Encounter Books.

Kaldor, M. (1999). *New and Old Wars: Organized Violence in a Global Era*. 1st edn. Cambridge: Polity Press.

——— (2007). *New and Old Wars: Organized Violence in a Global Era*. 2nd edn. Cambridge: Polity Press.

——— (2012). *New and Old Wars: Organized Violence in a Global Era*. 3rd edn. Cambridge: Polity Press.

——— (2013). 'In defence of New Wars'. *Stability: International Journal of Security and Development* 2(1), part 4. DOI: http://doi.org/10.5334/sta.at

Kallberg, J. (2018). 'Supremacy by accelerated warfare through the comprehension barrier and beyond: reaching the zero domain and cyberspace singularity'. *arXiv* 3(3).

Kania, E. B. (2019). 'Chinese military innovation in the AI revolution'. *The RUSI Journal* 164(5–6): 26–34.

Keenan, T. (2002). 'Publicity and indifference (Sarajevo on television)'. *PMLA* 117(1): 104–16.

Kennedy, L. (2015). 'Photojournalism and warfare in a postphotographic age'. *Photography & Culture* 8(2): 159–71.

Keller, U. (2002) *The Ultimate Spectacle: A Visual History of the Crimean War*. London: Routledge.

Khalili, L. (2020). *Sinews of War and Trade: Shipping and Capitalism in the Arabian Peninsula*. London: Verso Books.

Kilcullen, D. (2009). *Accidental Guerrilla: Fighting Small Wars amongst a Big One*. London: Hurst & Co.

——— (2013). *Out of the Mountains: The Coming Age of the Urban Guerrilla*. London: Hurst & Co.

——— (2020). *Dragons and Snakes: How the Rest Learned to Fight the West*. London: Hurst & Co.

King, A. (2013). *The Combat Soldier: Infantry Tactics and Cohesion in the Twentieth and Twenty-First Centuries*. Oxford: Oxford University Press.

Knox, M. and W. Murray (2001). 'Thinking about revolutions in warfare'. In M. Knox and M. Williamson (eds), *The Dynamics of Military Revolutions*. Cambridge: Cambridge University Press, pp. 1–14.

Knudsen, B. T. and C. Stage (2013). 'Online war memorials: YouTube as a democratic space of commemoration exemplified through video tributes to fallen Danish soldiers'. *Memory Studies* 6(4): 418–36.

Kollars, N. (2014). 'Military innovation's dialectic: gun trucks and rapid acquisition'. *Security Studies* 23(4): 787–813.

Kostyuk, N. and Y. M. Zhukov (2017). 'Invisible digital front: can cyber attacks shape battlefield events?' *Journal of Conflict Resolution* 63(2): 317–47.

Kott, A., A. Swami and B. West (2016). 'The internet of battle things'. *Computer* 49(12): 70–5.

Krepinevich, A. (1994). 'Cavalry to computer: the pattern of military revolutions'. *The National Interest* 37: 30–42.

Krieg, A. and J. Rickli (2019). *Surrogate Warfare: The Transformation of War in the Twenty-First Century*. Washington, DC: Georgetown University Press.

Kroker, A. (2014). *Exits to the Posthuman Future*. Cambridge: Polity Press.

Larson, M. S. (1979). *The Rise of Professionalism: A Sociological Analysis*. Berkeley: University of California.

Lasch, C. (1996). *The Revolt of the Elites and the Betrayal of Democracy*. New York: W. W. Norton.

Lembcke, J. (2000). *The Spitting Image: Myth, Memory and the Legacy of Vietnam*. New York: New York University Press.

Levy, S. (2020). *Facebook: The Inside Story*. London: Penguin.

Lewis, L. (2019). 'Resolving the battle over artificial intelligence in war'. *The RUSI Journal* 164(5–6): 62–71.

Lewis, M. (2014). *Flash Boys: A Wall Street Revolt*. London: Penguin.

Lind, W. S., K. Nightengale, J. F. Schmitt, J. W. Sutton and G. I. Wilson (1989). 'The changing face of war: into the fourth generation'. *Marines Corps Gazette*: 22–6.

Lindsay, J. R. (2020). *Information Technology and Military Power*. Ithaca, NY: Cornell University Press.

Lister, C. X. (2015). *The Syrian Jihad: Al-Qaeda, the Islamic State and the Evolution of an Insurgency*. London: Hurst & Co.

Livingstone, S. (2019). 'Audiences in an age of datafication: critical questions for media research'. *Television & New Media* 20(2): 170–83.

Lohaus, P. (2016). 'Special operations forces in the gray zone: an operational framework for using special operations forces in the space between war and peace'. *Special Operations Journal* 2(2): 75–91.

Lonsdale, D. (2003). *The Nature of War in the Information Age: Clausewitzian Future*. London: Frank Cass.

Lowenthal, D. (2012). 'The past made present'. *Historically Speaking* 13(4): 2–6.

Lucaites, J. L. and J. Simons (2017). 'Introduction: the paradox of war's in/visibility'. In J. Simons and J. L. Lucaites (eds), *In/visible War: The Culture of War in Twenty-First Century America*. New Brunswick, NJ: Rutgers University Press, pp. 1–24.

Lupion, M. (2018). 'The gray war of our time: information warfare and the Kremlin's weaponization of Russian-language digital news'. *The Journal of Slavic Military Studies* 31(3): 329–53.

Maçães, B. (2018). *Belt and Road: A Chinese World Order*. London: Hurst & Co.

MacKenzie, D. and G. Spinardi (1995). 'Tacit knowledge, weapons design, and the uninvention of nuclear weapons'. *American Journal of Sociology* 101(1): 44–99.

MacKinlay, J. C. G. (2009). *The Insurgent Archipelago*. London: Hurst & Co.

McDonald, J. (2017). *Enemies Known and Unknown: Targeted Killings in America's Transnational Wars*. London, Hurst.

McFate, S. (2019). *Goliath: Why the West Doesn't Win Wars; And What We Need to Do about It*. London: Penguin.

McLuhan, M. (1964). *Understanding Media: The Extensions of Man*. London: Routledge & Kegan Paul.

—— (1970). *Culture Is Our Business*. New York: McGraw-Hill.

McSorley, K. (2012). 'Helmetcams, militarized sensation and "somatic war"'. *Journal of War & Culture Studies* 5(1): 47–58.

Maltby, S. (2012). *Military Media Management: Negotiating the Frontline*. London: Routledge.

Marshall, S. L. A. (1947). *Men Against Fire: the Problem of Battle Command*. Norman, OK: University of Oklahoma Press.

Masco, J. (2014). *The Theatre of Operations: National Security Affect from the Cold War to the War on Terror*. Durham, NC: Duke University Press.

Merrin, W. (2014). *Media Studies 2.0*. London: Routledge.

—— (2018). *Digital War: A Critical Introduction*. London: Routledge.

Metahaven (2015). *Black Transparency: The Right to Know in the Age of Mass Surveillance*. Berlin: Sternberg Press.

Miller, T. (2020). *Violence*. London: Routledge.

Miskimmon, A., B. O'Loughlin and L. Roselle (2013). *Strategic Narratives: Communication Power and the New World Order*. London: Routledge.

Moss, M. and D. Thomas (2017). 'Overlapping temporalities: the judge, the historian and the citizen'. *Archives* 52(134): 51–66.

—— (2018). 'The accidental archive'. In C. Brown (ed.), *Archival Futures*. London: Facet Publishing, pp. 117–36.

Müller, J.-W. (ed.) (2002). *Memory and Power in Post-War Europe: Studies in the Presence of the Past*. Cambridge: Cambridge University Press.

Mumford, A. (2013). *Proxy Warfare*. Cambridge: Polity Press.

Münkler, H. (2005). *The New Wars*. Translated by Patrick Camiller. Cambridge: Polity Press.

Munoz, A. (2012). *U.S. Military Information Operations in Afghanistan: Effectiveness of Psychological Operations 2001–2010*. Arlington: RAND Corporation – National Defense Research Institute.

Nagl, J. (2005). *Learning to Eat Soup with a Knife: Counterinsurgency Lessons from Malaya to Vietnam*. Chicago: University of Chicago Press.

Nguyen, V. T. (2017). *Nothing Ever Dies: Vietnam and the Memory of War*. Cambridge, MA: Harvard University Press.

Nordin, A. H. M. and D. Öberg (2015). 'Targeting the ontology of war: from Clausewitz to Baudrillard'. *Millennium: Journal of International Studies* 43(2): 392–410.

Öberg, D. (2014). 'Forget Clausewitz'. *International Journal of Baudrillard Studies* 11(2).

O'Driscoll, C. (2019). 'No substitute for victory? Why just war theorists can't win'. *European Journal of International Relations* 26(1).

OECD (2016). 'States of fragility 2016'. Paris: OECD Publishing.

Oh, O., M. Agrawal and R. Rao (2011). 'Information control and terrorism: tracking the Mumbai terrorist attack through Twitter'. *Information Systems Frontiers* 13(1): 33–43.

O'Mara, M. (2020). *The Code: Silicon Valley and the Remaking of America*. New York: Penguin Random House.

Osinga, F. P. B. (2007). *Science, Strategy and War: The Strategic Theory of John Boyd*. New York: Routledge.

Owen, M. (2013). *No Easy Day: The Navy SEAL Mission That Killed Osama Bin Laden*. London: Penguin.

Paret, P. (2004). 'From ideal to ambiguity: Johannes von Müller, Clausewitz, and the people in arms'. *Journal of the History of Ideas* 65(1): 101–11.

Patrikarakos, D. (2017). *War in 140 Characters: How Social Media is Reshaping Conflict in the Twenty-First Century*. New York: Basic Books.

Paul, C. and M. Matthews (2016). 'The Russian "firehose of falsehood" propaganda model'. RAND Corporation.

Payne, K. (2021). *I, Warbot: The Dawn of Artificially Intelligent Conflict*. London: C. Hurst & Co.

Pomerantsev, P. (2015). *Nothing is True and Everything is Possible: Adventures in Modern Russia*. London: Faber & Faber.

—— (2019). *This is NOT Propaganda: Adventures in the War against Reality*. London: Faber & Faber.

Porch, D. (2013). *Counterinsurgency: Exposing the Myths of the New Way of War*. Cambridge: Cambridge University Press.

Porter, P. (2010). 'Why Britain doesn't do grand strategy'. *The RUSI Journal* 155(4): 6–12.

—— (2015). *The Global Village Myth: Distance, War and the Limits of Power*. London: Hurst & Co.

Postman, N. (1970). 'The reformed English curriculum'. In A. C. Eurich (ed.), *The Shape of the Future in American Secondary Education*. New York: Pitman Publishing Corporation, pp. 160–8.

Rabasa, A., R. D. Blackwill, P. Chalk, K. Cragin, C. C. Fair, B. A. Jackson, B. M. Jenkins, S. G. Jones, N. Shestak and A. Tellis (2009). *The Lessons of Mumbai*. Santa Monica: RAND Corporation.

Rasmussen, M. V. (2001). *The Acme of Skill: Clausewitz, Sun Tzu and the Revolution in Military Affairs*. Copenhagen: Dupi.

Renic, N. C. (2020). *Asymmetric Killing: Risk Avoidance, Just War, and the Warrior Ethos*. Oxford: Oxford University Press.

Rid, T. (2007). *War and Media Operations: The US Military and the Press from Vietnam to Iraq*. London: Routledge.

—— (2013). *Cyber War Will Not Take Place*. Oxford: Oxford University Press.

—— (2020). *Active Measures*. London: Profile Books.

Rid, T. and M. Hecker (2009). *War 2.0: Irregular Warfare in the Information Age*. Westport, CT: Praeger.

Rieff, D. (2016). *In Praise of Forgetting: Historical Memory and its Ironies*. New Haven: Yale University Press.

Roediger, H. L. and J. V. Wertsch (2008). 'Creating a new discipline of memory studies'. *Memory Studies* 1(1): 9–22.

Runia, E. (2014). *Moved by the Past: Discontinuity and Historical Mutation*. New York: Columbia University Press.

Russell, J. A. (2010). 'Innovation in war: counterinsurgency operations in Anbar and Ninewa provinces, Iraq, 2005–2007'. *Journal of Strategic Studies* 33(4): 595–624.

Ryan, M. (2019). *Full Spectrum Dominance: Irregular Warfare and the War on Terror*. Stanford: Stanford University Press.

Salber Phillips, M. (2004). 'History, memory and historical distance'. In P. Seixas (ed.), *Theorizing Historical Consciousness*. Toronto: University of Toronto Press, pp. 86–108.

Scharre, P. (2018). *Army of None: Autonomous Weapons and the Future of War*. New York: W. W. Norton.

Schrader, S. (2019). *Badges Without Borders: How Global Counterinsurgency Transformed American Policing*. Berkeley: University of California Press.

Scott, L. (2015). *The Four-Dimensional Human: Ways of Being in the Digital World*. London: William Heinemann.

Seib, P. (2021). *Information at War: Journalism, Disinformation, and Modern Warfare*. Cambridge: Polity Press.

Segal, A. (2018). 'When China rules the web: technology in service of the state'. *Foreign Affairs* 97(5): 10–18.

Shaw, M. (2003). 'Strategy and slaughter'. *Review of International Studies* 29(2): 269–77.

—— (2005). *The New Western Way of War*. Cambridge: Polity.

Sheikh, J. (2016). '"I just said it. The state": examining the motivations for Danish foreign fighting in Syria'. *Perspectives on Terrorism* 10(6): 59–67.

Silcock, B. W., C. B. Schwalbe and S. Keith (2008). '"Secret" casualties: images of injury and death in the Iraq War across media platforms'. *Journal of Mass Media Ethics* 23(1): 36–50.

Simons, J. and J. L. Lucaites (2017). *In/visible War: The Culture of War in Twenty-First Century America*. New Brunswick, NJ: Rutgers University Press.

Simpson, D. (2006). *9/11: The Culture of Commemoration*. Chicago: University of Chicago Press.

Singer, P. W. (2011). *Corporate Warriors: The Rise of the Privatized Military Industry*. Ithaca, NY: Cornell University Press.

—— (2018). *Like War: The Weaponization of Social Media*. New York: Houghton Mifflin Harcourt.

Smith, R. (2006). *The Utility of Force*. London: Penguin.

Smith, M. L. R. and D. Martin Jones (2015). *The Political Impossibility of Modern Counterinsurgency: Strategic Problems, Puzzles and Paradoxes*. New York: Columbia University Press.

Sontag, S. (2003). *Regarding the Pain of Others*. New York: Farrar, Straus and Giroux.

Southerton, C. (2020). 'Datafication'. In L. A. Schintler and C. L. McNeely (eds), *Encyclopedia of Big Data*. Cham: Springer International Publishing, pp. 1–4.

Srnicek, N. (2017). *Platform Capitalism*. Cambridge: Polity Press.

Stoker, D. (2019). *Why America Loses Wars: Limited War and US Strategy from the Korean War to the Present*. Cambridge, Cambridge University Press.

Stoker, D. and C. Whiteside (2020). 'Blurred lines: gray-zone conflict and hybrid war; two failures of American strategic thinking'. *Naval War College Review* 73(1): 1–37.

Strachan, H. (2005). 'The lost meaning of strategy'. *Survival* 47(3): 33–54.

Strauss, W. and N. Howe (1997). *The Fourth Turning: What the Cycles of History Tell Us About America's Next Rendezvous with Destiny*. New York: Crown Publishing.

Sturken, M. (1997). *Tangled Memories: The Vietnam War, the AIDS Epidemic, and the Politics of Remembering*. London: University California Press.

Suchman, L. (2020). 'Algorithmic warfare and the reinvention of accuracy'. *Critical Studies on Security* 8(2): 175–87.

Thomas, T. (2004). 'Russia's reflexive control theory and the military'. *The Journal of Slavic Military Studies* 17(2): 237–56.

Tilly, C. (1985). 'War making and state making as organized crime'. In P. Evans, D. Rueschemeyer and T. Skocpol (eds), *Bringing the State Back In*. Cambridge: Cambridge University Press, pp. 169–87.

Torgovnick, M. (2005). *The War Complex: World War II in Our Time*. Chicago: Chicago University Press.

Tufecki, Z. (2017). *Twitter and Tear Gas: The Power and Fragility of Networked Protest*. New Haven: Yale University Press.

Turner, F. (2008). *From Counterculture to Cyberculture: Stewart Brand, the Whole Earth Network, and the Rise of Digital Utopianism*. Chicago: University of Chicago Press.

Ucko, D. H. (2009). *The New Counterinsurgency Era: Transforming the U.S. military for modern wars*. Washington, D.C., Georgetown University Press.

Urban, M. (2010). *Task Force Black*. London: Little, Brown.

Virilio, P. (1989). *War and Cinema: The Logistics of Perception*. London: Verso.

—— (2002/1991). *Desert Screen: War at the Speed of Light*. London: Continuum.

—— (2009). *Grey Ecology*. Dresden: Atropos Press.

Wachter-Boettcher, S. (2018). 'Everything happens so much'. In D. Eggers (ed.), *McSweeney's Issue 54: The End of Trust*. San Francisco: McSweeney's Publishing, pp. 33–8.

Waldman, T. (2021). *Vicarious Warfare: American Strategy and the Illusion of War on the Cheap*. Bristol: Bristol University Press.

Wark, M. (1994). *Virtual Geography: Living with Global Media Events*. Bloomington: Indiana University Press.

Walker, R. M. (1992). *Inside/Outside: International Relations as Political Theory*. Cambridge: Cambridge University Press.

Webb, A. (2019). *The Big Nine: How the Tech Titans and Their Thinking Machines Could Warp Humanity?* New York: Public Affairs.

Weinberger, D. (2011). *Too Big to Know: Rethinking Knowledge Now That the Facts Aren't the Facts, Experts Are Everywhere, and the Smartest Person in the Room Is the Room*. New York: Basic Books.

Wertsch, J. V. (2002). *Voices of Collective Remembering*. Cambridge: Cambridge University Press.

Whiteside, C. (2020). 'Lying to win: the Islamic State media department's role in deception efforts'. *The RUSI Journal* 165(1): 130–41.

Whiteside, C., U. Rucem and D. Raineri (2019). 'Black ops: Islamic State and innovation in irregular warfare'. *Studies in Conflict & Terrorism* 44(1): 1190–217.

Williams, L. (2016). *Islamic State and the Mainstream Media*. Sydney: Lowy Institute.

Winter, C. (2016). 'Totalitarianism 101: the Islamic State's offline propaganda strategy'. Lawfare, 27 March.

Winter, J. (1995). *Sites of Memory, Sites of Mourning: The Great War in European Cultural History*. Cambridge: Cambridge University Press.

—— (2006). *Remembering War: The Great War Between Memory and History in the Twentieth Century*. New Haven: Yale University Press.

—— (2013). 'Human rights and European remembrance'. In U. Blacker, A. Etkind and J. Fedor (eds) *Memory and Theory in Eastern Europe*. New York: Palgrave Macmillan, pp. 43–58.

—— (2017). *War Beyond Words: Languages of Remembrance from the Great War to the Present*. Cambridge: Cambridge University Press.

Wirtz, J. J. (2017). 'Life in the "gray zone": observations for contemporary strategists'. *Defense & Security Analysis* 33(2): 106–14.

Wood, N. (1999). *Vectors of Memory: Legacies of Trauma in Postwar Europe*. Oxford: Berg.

Wu, T. (2016). *Attention Merchants: The Epic Struggle to Get Inside Our Heads*. London: Atlantic Books.

—— (2020). *The Curse of Bigness: How Corporate Giants Came to Rule the World*. London: Atlantic Books.

Yeo, G. (2017). 'Introduction to the series'. In D. Thomas, S. Fowler and V. Johnson. (eds), *The Silence of the Archive*. London: Facet Publishing, pp. ix–xi.

Zuboff, S. (2019). *The Age of Surveillance Capitalism: The Fight for a Human Future at the New Frontier of Power*. New York: PublicAffairs.

INDEX

Note: Page numbers followed by "*n*" refer to notes, "*f*" refer to figures.